BIOLOGICAL ASPECTS OF ANIMAL PRODUCTION

BIOLOGICAL ASPECTS OF ANIMAL PRODUCTION

Dr. Saurav Singh

Oxford Book Company
Jaipur ▪ New Delhi

Published By

Oxford Book Company

Regd. Off. "Bony Residency", Gate No. 2
Opp. Tilak Public School, Vishweswariya Nagar,
Gopalpura Road, Jaipur - 302018, India
Ph: 0141 - 2761280, Telefax: 2761381
email - oxfordbookjpr@gmail.com
website - oxfordbookcompany.com

Delhi Office: 102, 1st Floor, Satyam House,
4327/3, Ansari Road, Darya Ganj,
New Delhi - 110002, India
Ph: 011-45652440

Distributed By

KEYA INTERNATIONAL PRESS LLP
E-mail: keyainternationalpressllp@gmail.com

Published in India, 2025

ISBN : 9789391096700

Typeset by Shivangi Computers, Jaipur

Printed and bound in India by Thomson Press Ltd.,
New Delhi

Contents

Preface

The art of animal breeding is very much older than the science of genetics, and even today the geneticist cannot help the animal breeder nearly as much as he can the breeder of plants. But fortunately the contribution of the geneticist is steadily increasing. In many spheres progress in animal breeding has almost stopped and further advances will depend on the application of the principles of genetics. Let us look at a few cases. The most important meat animal in is the lamb. Now meat consists largely of muscle and fat, two tissues which develop relatively late in most animals. To produce an attractive fat lamb it is necessary that these tissues should develop earlier than usual. Fortunately the characters responsible for early maturity in the Southdown sheep are so dominant that a Southdown ram may be crossed with a ewe of any breed and the lamb will mature early provided it receives adequate milk from the ewe. The Romney Marsh ewe nearly always does provide enough milk for one lamb and generally enough for two. Consequently lamb is famous for its quality. But there still remains a major problem for the geneticist. The fat lamb breeder seldom produces more than 100% of fat lambs from his ewes. Using a different breed in England, 150% has been achieved. The breed used in England would probably not suit New Zealand conditions, but we badly need that extra 50% of lambs to help consume the spring and summer flush of grass much of which at present, is often wasted. Twinning in sheep is not very strongly inherited; it may be a recessive character. It needs to be bred for in the Romney Marsh without reducing milk and wool production–a nice problem in animal genetics.

An even more baffling problem faces the breeder of dairy cattle. Many dairy farmers derive their income almost entirely from the sale of butterfat. They therefore desire cows which will produce large volumes of milk with a high concentration of butterfat, characters which are inherited separately. The inheritance of high production of milk appears to be very complex. Moreover, this character does not find expression in cows until they are two years old and it remains completely dormant in bulls. No method has been devised for assessing the genotype of bulls for butterfat production based on examination of their phenotype, and the position in regard to cows is very little better. In such cases it is necessary to resort to the practice of sire surveying in which a bull's genotype is measured from the production of his first group of daughters. Though reasonably effective this method is very cumbersome and slow and it seldom reveals a bull's worth before more than half of his breeding life is passed. However, the consistent uses of the best progeny tested sires in stud herds by means of artificial insemination should result in a gradual concentration of the desirable genes. Concurrent with this practice should go an intensive search for measurable manifestations in the bull of those genes which at present can only be estimated after they have been transmitted to his daughters. To this end a study is being made of some thirty erythrocyte antigens which are inherited as simple dominants. Should the genes for any of these reside on the same chromosomes as genes responsible for high butterfat production, it might become possible to assay the productive potentialities of both bulls and heifers at an early age. American workers have suggested that a heifer's capacity for lactation may be assessed from the degree of development of mammary tissue which occurs at an early age. This would be extremely valuable as it would save feed at present wasted on low producers during the first two years of their lives. Occasionally an undesirable character is found and it is necessary to determine its mode of inheritance so that it may be eliminated. An interesting example is a photosensitivity which occurs in some Southdown sheep.

In essence the book could be of vital significance for sustainable animal production and feeding technology.

—Dr. Saurav Singh

Chapter 1

Ingredients for Fish Feed Manufacture

The abundance of feed grains in the U.S.A. often obscures the importance of industrial sources of raw materials for the domestic feed industry. Published data from research in poultry and swine nutrition show corn or wheat as the predominant dietary component and soybean meal or fish meal as the only dietary protein source in test diets. Often a test ration would consist of just a cereal grain and soybean meal plus mineral and vitamin supplements.

Increasing competition in the feed industry plus rising cost of traditional feed ingredients in recent years are forcing feed manufacturers to increase the use of less expensive feed ingredients such as by-products from the milling and food processing industries. Feed formulations have been aided by more complete and reliable information of the nutritive value of these products as well as by the development of linear programming techniques for least cost diet formulation based on a large selection of ingredients.

Some of the more commonly used industrial by-products currently used for fish diet manufacture in the U.S.A. are listed below.

Mill By-products

Wheat Germ Meal

Wheat germ meal consists chiefly of the germs of the wheat berry, with some bran and middlings. A rather wide variety of wheat germ grades are produced, depending on the area and regional demand. The quality of wheat germ meal is influenced by the presence of screenings and the level, of other wheat products, which can lower the fat or protein level. Excessive storage time may result in rancidity, due to the oxidation of the rather high level of fat. Generally, I wheat germ meal may have a crude protein level of 25-30 percent; crude fat, 7-12 percent; and crude fibre, 3-6 percent.

Wheat Red Dog

Wheat red dog consists of offal from the tail of the mill, together with some fine particles of wheat bran, wheat germ and wheat flour. Wheat red dog is a fine, floury, light-coloured feed ingredient. The colour may range from creamy white to light brown or light red, depending on the type of wheat being milled. Wheat red dog can act as a pellet binder, as well as contributor of protein, carbohydrates, minerals and vitamins to the feed. Wheat red dog may have a crude protein level of 15.5-17.5 percent; crude fat, 3.5-4.5 percent; and crude fibre, 2.8-4.0 percent.

Wheat Bran

Wheat bran is the coarse outer covering of the wheat berry. Sometimes screenings are ground and added to the bran. Wheat bran contributes bulk to ruminant feeds and is a source of carbohydrates, protein, minerals and vitamins. Generally, wheat bran will have a crude protein level of 14-17 percent; crude fat, 3-4.5 percent; and crude fibre, 10.5-12 percent.

Wheat Middlings

Wheat middlings consist of fine particles of wheat bran, wheat shorts, wheat germ, wheat flour and some of the offal from the tail of the mill. The quality of middlings may be influenced by the amount of screenings added and the fineness of grind. A moisture level over 14 percent may cause problems, especially in hot, humid climates. Wheat middlings have a crude protein, level of 15-17 percent; crude fat, 3-4 percent; and crude fibre, 8.5-9.5 percent.

Wheat Mill Run

Wheat mill run consists of wheat bran, fine particles of wheat bran, wheat shorts, wheat germ, wheat flour and the offal from the tail of the mill. Ground run of the mill screenings are normally added. Wheat mill run is available in the west, but not normally found in other areas of the country where mill run is separated into bran, middlings and red dog. Wheat mill run may have a crude protein level of 14-17 percent; crude fat, 3-4 percent and crude fibre, 8.5-9.5 percent.

Corn Gluten Feed

Corn gluten feed is that part of commercial shelled corn that remains after the extraction of the larger portion of the starch, gluten and germ by the process employed in the wet milling manufacture of corn starch or syrup. Generally, corn gluten feed will contain a crude protein level of 21-22 percent; crude fat, 1-2.5 percent; and crude fibre, 8-10 percent.

Corn Gluten Meal

Corn gluten meal is the dried residue from corn after the larger part of the starch and germ have been removed and the bran separated by the process employed in the wet milling manufacture of corn starch or syrup, or by enzymatic treatment of the endosperm. Corn gluten meal may contain fermented corn extractives and/or corn germ meal. The crude protein level may range from 41-43 percent; crude fat, 1-3 percent; crude fibre, 4-6 percent; and xanthophyll, 140-220 mg/kg.

Rice Groats

Rice groats are the ground rice kernel, without the bran. Rice groats contain approximately a crude protein level of 8.6 percent; crude fat, 1.2 percent; and crude fibre, 1.0 percent.

Rice Bran with Germ

Rice bran with germ is primarily the pericarp or bran layer and the germ portions of the rice kernel. The crude protein level is approximately 13.3 percent; crude fat, 15.0 percent; and crude fibre, 11.0 percent.

Rice Polishings

Rice polishings are the outer portions of the rice kernel produced by the process used in producing polished white rice. Rice polishings have a crude protein level of 11.8 percent; crude fat, 13.2 percent; and crude fibre, 3.0 percent.

Oat Groats

Oat groats are the cleaned oats with their hulls removed. Generally, oat groats will contain a crude protein level of 16 percent; crude fat, 6 percent', and crude fibre, 2 percent.

Oil Extractives

Soybean Meals

Soybean meal is produced in two major protein levels by different processes. Soybean meal must be heated (toasted) sufficiently to destroy, the trypsin inhibitor. Forty-four percent soybean meal is usually mechanically extracted to produce a meal of 44 percent crude protein; crude fat, 4.7 percent; and crude fibre, 6.0 percent. Forty-eight percent soybean meal is dehulled and solvent extracted to yield meal with a crude protein I level of 48 percent; crude fat, 0.9 percent; and crude fibre, 2.8 percent.

Cottonsead Meals

Cottonseed meal is produced in three protein levels. Cottonseed meals are classified as low gossypol if they contain less than 0.04 percent free gossypol. Forty-one percent cottonseed meal has a crude protein. level of 41 percent; crude fat, 2 percent; and crude fibre, 12 percent. Forty-eight percent cottonseed meal has a crude protein level of 48 percent; crude fat, 1 percent; and crude fibre, 8 percent. Although all three cottonseed meals are solvent extracted, the lower protein meals may contain some hulls.

Peanut Meal

Peanut meal is obtained by grinding the cake, chips or flakes obtained by removal of most of the oil from peanut kernels by a mechanical or solvent extraction process. Peanut meal will have a crude protein level of about 46 percent; crude fat, 1-7 percent; and crude fibre, 10-13 percent.

Animal By-products

Hydrolysed Poultry Feathers

Feathers from poultry are collected, ground and hydrolysed under pressure. During processing the cystine linkage is broken, increasing the value of the meal. Not less than 75 percent of the crude protein must be pepsin digestible. Generally, hydrolysed poultry feathers will range in crude protein from 80 to 85 percent; crude fat, 2.5 percent; crude fibre, 1.5 percent; phosphorus, 0.75 percent; and ash about 3 percent.

Meat and Bone Meal

Meat and bone meal is the dry rendered product derived from mammalian tissue, exclusive of hair, hoof, horn, manure and paunch contents. The relatively high fat content of meat and bone meal apparently helps protect the lysine content during rendering. Meat and bone meal generally contain a crude protein of 50 percent; fat, 8-11 percent; crude fibre, 3 percent phosphorus, 4.4-5 percent; calcium, 8.8-10 percent; and ash, about 30 percent.

Fish Meals

Fish meals are manufactured by cooking fish, pressing to remove water and oil, and then drying. Fish meals are often made from a single species of fish, e.g., herring meal, but meals containing more than one species, such as bottom fish meal, are also sold. Generally, the single species meals are more uniform in quality. Some typical analyses follow:

	Crude protein (%)	*Crude fat (%)*	*Crude fibre (%)*	*Sodium (%)*
Herring meal	70	7.5	1	0.5
Menhaden meal	60	7.5	-	0.3
Anchovy meal	64	6	1	0.8
Carp meal, whole	75	9.5	-	-
Shrimp meal	36-48	3	11	-
Crab meal	30	1.7	11	-
Bottom fish meal	60	-	-	-

Blood Meal

Blood meal is produced from clean fresh animal blood, exclusive of all extraneous material such as hair, stomach contents, etc. Blood meal may be dried by several processes, but most often by spray drying. Spray dried blood meal has approximately a crude protein level of 85 percent; crude fat, 0.5-3 percent; crude fibre, 2.5 percent; ash, 6 percent; and lysine, 9-11 percent, with an availability of 80-90 percent.

Miscellaneous

Corn Distillers Dried Grain with Solubles: This product is obtained by removal of ethyl alcohol by distillation from the yeast fermentation of corn or corn mixture by condensing and drying at least 75 percent of the solids of the resultant whole stillage. Corn distillers dried grain with solubles has a crude protein level of 26-27 percent; crude fat, 7-8 percent; and crude fibre, 8.5-9.5 percent.

Brewers Yeast

Brewers dried yeast is the dried, sterilised, unextracted yeast (Saccharomyces) resulting as a by-product from the brewing of beer and ale. Brewers dried yeast has a crude protein level of 45 percent; crude fat, 1 percent; and crude fibre, 2.7 percent.

Brewers Dried Grains

Brewers dried grain is the extracted dried residue of barley malt alone or in mixture with other cereal grain or grain products resulting from the manufacture of work or beer and may contain pulverised spent hops in an amount not to exceed 3 percent. Brewers dried grains may have a crude protein level of 24-30 percent; crude fat, 4.5-8.5 percent; and crude fibre, 10.5-16 percent.

Dried Whey

Dried whey is the residue obtained by drying whey, a by-product of cheese manufacturing. It contains at least 65 percent lactose; crude protein, 13 percent; about 0.8 percent crude fat; and no crude fibre.

Antioxidants in Compounded Feeds

Just as oxidation can cause rust and deterioration in metals, a similar type of oxidation occurs in feeds and feedstuffs, resulting in rancidity of fats, destruction of vitamins A, D, and E, pigmenters (carotenoids) and amino acids with resultant lowered biological energy values for the diet. If this destruction is allowed to proceed unchecked in a feed, or even in a single ingredient, a lowered feed consumption may result in disastrous nutrient deficiencies. Researchers in a number of fields have been studying the various problems resulting from uncontrolled oxidation and have been developing ways of bringing these oxidation processes under control. It is noted that oxidative rancidity or lipid peroxidation, as contrasted with hydrolytic rancidity, results in a serious decrease in the energy value of a fat or oil.

Undesirable oxidation in feeds may be combatted in several ways. Care should be used to make certain that the ingredients included in the feeds provide adequate margins of safety of vitamins A, E, and other natural antioxidants; e.g., lecithin. The use of unstable fats and oils or other pro-oxidants in the feed should be minimised whenever possible.

Criteria for Selecting Feed Antioxidants

Antioxidants have been used in commercial fish foods in the USA for over 20 years. Although hundreds of chemicals have been tested, only a few have shown the qualifications necessary to make them suitable for use in preventing undesirable oxidations in feedstuffs, in finished feeds, and in the guts and carcasses of animals. In order for an antioxidant to be useful in animal feeding, it must have the following qualifications:

(a) it must be effective in preserving animal and vegetable fats, vitamins, and other feed qualities subject to oxidative destruction;

(b) it must be non-toxic to man and to farm animals (i.e., chickens, swine, fish, etc.);

(c) it should be effective at very low concentrations; and

(d) it must be low enough in cost to be economically practical.

Commonly Used Feed Antioxidants

Of the chemical compounds that have been investigated thus far, three have been found to be outstandingly effective antioxidants for feeds and feed ingredients and can be used both efficiently and economically. They are:

(a) Ethoxyquin (generic term: 1,2-dihydro-6-ethoxy-2,2,4-trimethylquinoline)

(b) BHA (butylated hydroxyanisole);

(c) BHT (butylated hydroxytoluene)

Ethoxyquin, however, has been demonstrated to be the most efficaceous, followed closely by BHT and BHA.

With the advent of rations containing a high level of animal and vegetable fats, the requirement for antioxidant protection has become very apparent. The majority of studies over the last few years has focused on ethoxyquin as a preservative or antioxidant. Other chemical preservatives are: ascorbic acid, propionic acid, benzoic acid, citric acid and their various salts. There are technological problems (i.e., moisture level, etc.) associated with the use of these preservatives. Economics, however, remain the most important consideration which limit their use in fish foods.

Ascorbic acid Ascorbyl palmitate Benzoic acid BHABHTCalcium ascorbate Calcium propionate Calcium sorbate Citrate acid Dilauryl thiodipropionate Distearyl thiodipropionate Erythorbic acid Ethoxyquin Formic acid Methylparaben Potassium bisulphite Potassium metabisulphite Potassium sorbate Propionic acid Propul gallate Propul paraben Resin guaiae Sodium ascorbate Sodium benzoate Sodium bisulphite Sodium metabisulphite Sodium nitrite Sodium propionate Sodium sorbate Sodium sulphite Sorbic acid Stannous chloride Sulphur dioxide THBP - Trihydroxy-butyrophenone TBHQ - Tertiary-butylhydroquinone Thiodipinic acid Tocopherols

Functional Effects of Antioxidants

The functional effects of the antioxidants as they affect nutrients and the animals' utilisation of them can be summarised as follows:

Nutrient Deficiency Prevention

Dietary deficiencies of vitamins A and E seem to be ameliorated in certain circumstances and ethoxyquin promotes higher levels of vitamin A storage in the liver. Repletion/deletion experiments show that in both monogastric and ruminant animals, a diet containing an anti-oxidant protects fat soluble vitamins throughout ingestion and metabolism. The important benefit of antioxidants most probably lies in their conservation of essential nutrients and their improved utilisation by the animal. Altogether too often, it is the practice to use levels of vitamin E far above the animals' nutrient requirement and the result is economically unfavourable. It has been shown in diets designed for chicken and turkey breeders that ethoxyquin has a vitamin E sparing effect.

Prevention of Rancid Oxidation of Fats

In lipid peroxidation, the unsaturated fatty acids undergo a loss of hydrogen, resulting in the formation of a free radical at the site of unsaturation.

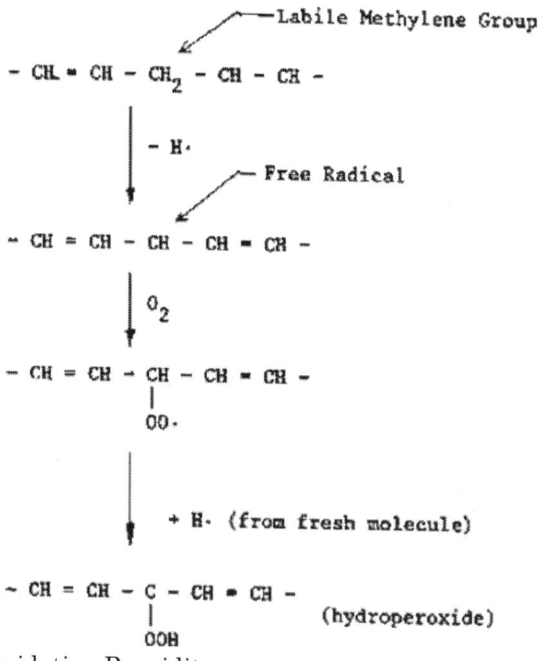

Figure 1: Peroxidative Rancidity

If the feed material in which this reaction is taking place does not contain vitamin E or some other effective antioxidant, the free radical is quickly converted to a fatty acid peroxide free radical and finally to a fatty acid hydroperoxide.

An antioxidant can block this peroxidation by supplying a hydrogen in the first free radical formed, thereby reconverting it to the original fatty acid. If the hydroperoxides are allowed to form, they continue to decompose by breaking down into a variety of aldehydes and ketones.

Antioxidants prevent oxidative losses of vitamins A and E and pigmenters (oxy- and keto-cerotenoids) in stored mixed feeds. Antioxidants stabilise critical oxidation-susceptible nutrients that are naturally present in a fish feed composed of several feedstuffs so that losses are minimal from mixing and storing.

If pigmenting substances are used, the anti-oxidants are definitely needed. The benefits of adequate, consistent use span all facets of fish production which include the processing and handling of feedstuffs, formulation, and fish cultural practices.

Table 1: Some Effects of Ethoxyquin in Feeds

Sample	*Vitamin A activity, I.U.**		*Change (%)*
Feed A (no antioxidant added)	Initial 150 325		
	Final 77 800		- 48.2
Feed B (150 ppm Ethoxyquin)	Initial 151 366		
	Final 115 633		- 23.6
Sample	*Extractable fat**	*Iodine No.**	*ME (cal/lb)**
Fish meal (unstabilized)	7.0	109	1 149
Fish meal (stabilized)	9.4	178	1 461
Hatchability of eggs from hens (12 weeks, 7% linoleic acid diet)			
Supplementation		*Egg hatchability (%)*	
None		55.50	
Ethoxyquin, 125 ppm		78.96	
Vitamin E, 16 mg/lb		82.11	
Ethoxyquin, 125 ppm and Vitamin E, 16 mg/lb		79.15	

Level of Antioxidant Usage in Feed

The U.S. Food and Drug Administration permits the following levels of antioxidant in finished feed:

(a) ethoxyquin (1,2 dihydro-6-ethoxy-2,2,4- trimethy quinoline): 150 ppm,

(b) BHT (butylated hydroxytoluene): 200 ppm, and

(c) BHA (butylated hydroxyanisole): 200 ppm.

Other Effects of Antioxidants in Feed

It was recently found that BHT, which is widely used as a poultry food additive, prevented mortality in chickens exposed to virulent Newcastle disease virus but prevented an antibody response when chickens were exposed to a virulent Newcastle disease virus. This in vivo effect of BHT in rations containing 100-150 ppm suggests a possible explanation for vaccination failure sometimes encountered with live vaccines and possibly a new direction in the development of antiviral therapy. Researchers are now trying to define the mechanism of the apparent antiviral action of BHT and extending their attention to other viruses using both BHT and other antioxidants.

Adventitious Toxins in Feeds

Toxic materials affecting fish feed ingredients are my cotoxins, toxic products in plant materials, bacteria, residues of pesticides, drugs, and agricultural chemicals. Since my cotoxins are covered in another chapter, the discussion here will revolve around the other adventitious toxins.

Chemical Contaminants

Work at the Columbia Pesticide Laboratory of the U.S. Fish and Wildlife Service shows that the most common dietary contaminants have been the organo-chlorine pesticides DDT, DDE, dieldrin, endrin, and industrial chemicals such as polychlorinated biphenyls (PCB), phthalate esters, and hexachlorobenzene (HCB). Most, but not all, residues are due to the presence of contaminated fish oils and/or meal in the finished diets.

Pesticides and Chlorinated Hydrocarbons

The effects of the chlorinated hydrocarbons have been reasonably well documented in the published literature. Detrimental effects on egg hatchability and survival of young fish are quite striking. Toxaphene has been shown to interfere with vitamin C metabolism which affects collagen synthesis and ultimately causes skeletal problems. Dieldrin affects the phenylalanine hydroxylase enzyme which is involved in amino acid metabolism and glutamine synthetase which is involved in protein metabolism. The inhibition of phenylalanine hydroxylase increased the level of blood phenylketo acids, a condition similar to the genetic metabolic disorder phenylketonuria (PKU) that is associated with mental retardation in humans; inhibition of glutamine synthetase might interfere with ammonia detoxification which would be physiologically deleterious. A stimulation of thyroid activity has been associated with DDT and its analogues and endrin has been implicated in the inhibited mobilisation of liver glycogen. It would appear that serious attention should be focused on pesticides and other industrial contaminants in dietary feedstuffs and finished fish feeds.

The Columbia National Fishery Laboratory has made recommendations that fish oils used at 3-5% of the diet should contain no more than 2 ppm and fish meals no more than 0.1 ppm of organochlorine contaminants, including PCBs. In addition, the U.S. Food and Drug Administration (FDA) regulations state that the maximum amount of PCBs permitted in animal feed for food producing animals is 0.2 ppm, and the amount permitted in animal feed components is 2.0 ppm.

Seed Dressings

Another serious group of contaminants of the ration that may gain entry anywhere from the ingredient supplier to the fish feeder include the newer seed disinfectants and fungicides, and certain drugs and chemicals added to feeds. Most seed treatments include organic mercury compounds which are very chemically reactive and readily absorbed. The compounds are highly toxic and the amounts absorbed in light, accidental, contacts accumulate and build up their effects on fish.

Cross Contaminants

Cross contamination occurs when trace amounts of a drug or chemical used as a feed additive are incorporated into a feed which is not supposed to contain the drug. The feed usually affected is the batch immediately following a medicated feed which legally contains the drug in question.

Measurable amounts of the drug-containing feed may remain in the production system and become intermingled with other feed being manufactured. It may occur in one piece of equipment or it may be a combination of residues throughout the entire system. To discover the source, all equipment from the point of addition of the drug to the point of shipping must be checked.

Although there are some partial remedies that can be taken in scheduling feed production or with special procedures between feed mixings, experience has shown they are less than adequate answers to the problem. The safest procedures would involve having a completely separate production line or, better yet, a feed manufacturing facility devoted exclusively to fish feed.

Other Sources of Chemical Contaminants

Boiler water additives, hydraulic fluids, lubricants (on pellet mill rolls and other processing equipment), fumigants, defaming agents, and coating materials are examples of materials that could become feed additives during feed manufacture. Misuse of these materials could conceivably lead to human loss as well as fish mortality.

Biological Toxins

Salmonella: A very serious source of contamination usually overlooked is that of rodent contamination of stored ingredients. There are approximately 1200 serotypes of salmonella organisms that inhabit a great variety of warm and cold blooded animals, including fish. Of these, approximately 80 serotypes have been definitely associated with diseases. Animal by-product derivatives are the worst offenders and the primary source of salmonella contamination in mixed feeds. Fish meals represent the next highest source of salmonella contamination. Attention has thus far focused on feeds being the means of perpetuating the salmonella cycle (rat to faeces to by-products to feed to fish). Attention also should be given to the role of soil, water, contaminated feeding equipment, and wild birds and terrestial animals as contributing to the presence of salmonella organisms in a given environment.

Antinutritional Factors

Any discussion on adventitious toxins has to include the antinutritional factors in plant proteins. Nearly all sources of plant protein possess associated factors which must be eliminated by special processing techniques to make them of maximum nutritional value. For example, for over fifty years, soybeans have been known to be improved by heat treatment.

Soybeans

Factors present in raw soybeans can markedly affect the intestinal tract of animals and influence the digestion and utilisation of many nutrients. Inclusion of small amounts of soybean meal will depress growth rate, increase pancreas size, decrease fat absorption, and metabolisable energy of the diet of fish. In addition, it has also been found that the feeding of raw soybeans causes the gall-bladder to contract, increases excretion of bile acids, lowers intestinal proteolytic activity, and affects methionine metabolism.

Raw soybeans also contain protein which cause agglutination of red blood cells in vitro. It appears that the haemagglutinin (lectins) of raw soybeans are responsible for some of the decrease in efficiency of feed utilisation.

Apparently these lectins are all capable of causing release of intestinal membrane-bound lipases and amylases, causing these digestive enzymes to be eliminated in the faeces, thereby reducing their digestive capabilities. Fortunately, all of these factors can be destroyed or reduced to minimal levels by proper heat treatment.

Other Vegetable Protein Sources

Groundnuts contain trypsin inhibitor activity which is not destroyed by heat treatment. This is thought to be due to the fact that the inhibitor activity resides in the tannins which are a part of the groundnut skin cottonseed contains gossypol and fatty acids with a cyclopropene ring which causes nutritional difficulties. Linseed meal contains an antipyridoxine factor and a cyanogenic glycoside.

Rapeseed meal contains glucosides which when hydrolysed yield a goitrogen. Most all plant protein sources contain phytic acid which can interfere with both mineral and protein availability. Particular attention must be paid to mineral supplementation when formulating fish diets containing fairly high amounts of plant protein.

Even though the temperature which destroys heat labile growth inhibitors is high, minimum protein damage is involved at these high temperatures provided the optimum peak of temperature is reached for a very brief period of time. The brevity of the period at peak temperature is more important than any other factor in obtaining! optimum protein quality.

Unconventional Feed Ingredients for Fish Feed

The rapid world-wide expansion of aquaculture and livestock production strongly indicates that a crisis will be precipitated in the livestock and aquaculture feed industries in the near future. Food for humans is not included in this consideration because, generally speaking, livestock, fish and humans can all eat the same basic food commodities and, in emergencies or times of scarcity, feedstuffs are eaten first by humans. The consumption of cultured livestock, fish, etc., is linked to human affluence. For example, per capita meat consumption in the United States before 1941 was about 50 lb; it is now over 150 lb. Similar percentage increases have occurred in Japan and the other industrialised countries of the world. Fish are probably the most efficient converters of feed to flesh, requiring from 2 to 4 lb of basic feedstuffs to produce 1 lb of fish. In contrast to land animals, however, fish are fastidious eaters in that they require higher levels of dietary protein. In addition, the amino acid requirements to promote rapid growth of fish appear to be more rigid than for land animals. As an extreme example, the protein in the diets of fish such as eels and yellow tail need to be almost entirely of animal origin, while the protein requirements of ruminants can be partially satisfied from non-protein nitrogen sources such as urea and biuret.

This chapter focuses primarily on the development work now being conducted on alternative or unconventional feed ingredients as suitable replacements for fish meal in artificial diets for intensive aquaculture.

Problems Associated with Unconventional Feedstuffs

Faced with the food supply problem for cultured fish, nutritionists have done more work evaluating alternate protein sources in aquaculture diets during the last 7 years than during the previous 50 years. A review of the literature shows that practically no potential feed material has been ignored.

Feedstuffs of vegetable origin are as a whole lower in protein content when compared with those of animal origin. In addition, the presence of high amounts of carbohydrates, fibre, and other organic molecules such as glucosides, phytates, and cyclopropenes in these sources present the nutritionist with problems that are generally not encountered with sources of animal origin. Despite these problems, practically all the commercialised feedstuffs listed are being used to some extent in commercial aquaculture diets. None, however, perform as well as fish meal on a protein equivalent basis. The problems relating to their usage in fish diets will be discussed in greater detail in the next section.

Of the non-commercial items, only single cell protein (SCP), krill meal, and possibly leaf protein concentrate have a chance of becoming commercial within three to five years. Because SCP are synthetic products and because the biomass of krill is so huge (the latter five times that of all landed fish), the volume of either could dwarf everything listed in except soyabean meal.

Formulation Problems

As previously mentioned, the nutritive requirements of fish are such that they probably offer less flexibility in diet formulation than do those of most land animals. First of all, several species of fish selected for culture are almost pure carnivores, requiring a diet of high protein content. These fish have very poor utilisation of carbohydrate as an energy source, and some evidence is developing that the inclusion of certain types of carbohydrates in the diet is, in fact, detrimental. Hence, it has been recently reported that soyabean meal extracted with alcohol performed better in rations than unextracted meal. This improvement has been attributed to the removal of low molecular weight carbohydrates by the alcohol. The subject of carbohydrates in soyabean meal will be discussed later.

Table 2: Alternate Sources of Protein that are Being Evaluated or have Potential as Partial or Whole Replacement for Fish Meal in Aquaculture Diets

Commercialized		***Not commercialized***	
Vegetable	***Animal***		
Soy meal	Poultry byproducts	Insect larvae	
Rapeseed meal	Feather meal	Single cell protein	
Sunflower meal	Shrimp and crab meal	Grasses	
Oat groats	Blood flour	Leaf protein	
Cottonseed meal	Fish silage	Vegetable silage	
Wheat middlings	Meat meal	Zooplankton (krill, etc.)	
		Recycled wastes	
			Yeast
			Phytoplankton
			Bacteria
			Algae
			Higher plants
Protein (range), %			
15-50	50-85	4-85	

Feeds tuffs contain protein in amounts ranging from about 15 to 50 percent. This falls within the range of the protein requirements for optimum growth of several species of fish.

However, proteins derived from vegetable sources are somewhat deficient in several key amino acids such as lysine, methionine, and tryptophane.

This, of course, is not the only problem when unconventional sources of protein are used. A fish diet must provide a suitable energy source and be in proper balance with respect to:

(i) proteins,

(ii) minerals,

(iii) lipids,

(iv) carbohydrates, and

(v) vitamins and growth factors.

Then, there are the factors that relate to the feeds tuffs comprising the diet. These include:

(a) *Composition:* The commodity must have a composition that allows it to be compounded into a balanced diet. For example, because of the high water content, it would be difficult to compound potential feeds tuffs such as sea plants, algae, etc., into diets.

(b) *Physical form:* Many feedstuffs must be modified for proper formulation into diets. A common example in salmon diets is the lint that remains in cottonseed meal. It readily plugs the dies when small diametre feeds are to be prepared.

(c) *Palatability:* Several potential feedstuffs contain compounds that are offensive to the olfatory receptors of the fish.

(d) *Factors affecting bio-availability of nutrients:* (e) stability during storage

This primarily relates to the vitamin stability and also the stability of the lipid portion that may oxidise in either dry or frozen storage.

(f) Toxic factors.

Of all the feedstuffs investigated as alternatives to fishmeal in aquaculture diets, soyabean meal has received the most attention.

Table 3: Some Compounds Occurring in Feedstuffs that are Known and/or Suspected of Causing Physiological Abnormalities or Otherwise Impairing the Growth of Fish

Compounds	***Found in***
Glycosides	Grass and leaves
Phytates	All plant foods tuffs
Mycotoxins (aflatoxin)	Cereal-based meals not naturally occurring but produced by microorganisms
Cyclopropenoid fatty acids	Cottonseed oil and meal
Trypsin inhibitors	Soy and rapeseed meal
Mimosine	Leaves (Leucaena leucocephala)
Glucosinolates	Rapeseed meal
Haemoglutinins	Soyabean meal
Plant phenolics	
-Gossypol	Cottonseed meal
-Tannins	Rapeseed meal
Oxidised and polymerised lipids	Fish meal; poultry byproducts, krill meal
Histamine and putrescine	Fish meal, primarily tuna
Nitrosamines	Fish meal

Chapter 2

Selected Unconventional Ingredients for Fish Feed

Soyabean Meal

With regard to its composition, soyabean meal appears to be a reasonably good feed component for aquaculture diets. It contains about 47-50 percent protein, 5-6 percent ash, 1 percent lipid and about 40 percent carbohydrates. It has a lysine content that approaches that of fishmeal, but for most aquaculture diets it is deficient in the sulphur-containing amino acids and in tryptophan. Because of these amino acid deficiencies, soyabean meal cannot be used as the only source of protein and is generally compounded with other feed-stuffs when it is used in aquaculture rations.

Although the inclusion of soyabean meal in fish diets presents no manufacturing problems, problems relating to palatability and availability of nutrients have been encountered. There is abundant literature discussing the deficiencies and merits of this feedstuff. Several investigators have demonstrated that heating soyabean meal rather severely not only increases its acceptability to fish, but also improves the availability of nutrients.

This is first accomplished by deactivating trypsin inhibitors, and second by denaturing the protein and thus making it more digestible, and third by detoxifying natural toxicants. Several studies, however, indicate that when soyabean meal is heated, its performance is still below the expected considering its amino acid composition. Recent studies with catfish, for example, show that even when soyabean meal was fortified with lysine, cystine, and methionine, growth was significantly lower than when menhaden meal was used as the source of protein. It was concluded, among other things, that soyabean meal must contain some anti-growth factors or that catfish do not utilise free amino acids. Other workers, however, have shown that the latter conclusion might not be true, since the carp and salmon can utilise free amino acids.

Work done at the Northwest and Alaska Fisheries Centre has shown that when soyabean meal substituted for fishmeal in Oregon moist pellet (OMP) type diets, growth is significantly reduced when these diets are fed to rainbow trout. It is believed that part of the problem is due to the phytate content of soyabeans.

Chemically, phytic acid is the hexametaphosphate of myoinositol. In plant material it is supposed to occur as the Ca-Mg salt of the hexametaphosphate. The Merck Index states that there are five calcium (Ca) ions and one magnesium (Mg) ion per mole. This may vary with plant species as well as with the maturity of the plant seed. Its chemical structure is still open to question because:

(a) theoretically, it can have several isometric forms (i.e., nine stereoisomers and two mesoforms),

(b) it can form crosslinks between phosphate groups,

(c) the mechanism for Ca and Mg binding remains uncertain, and

(d) its ability to bind other ions depends on how much Ca is present.

In plants, phytic acid occurs as the Ca-Mg salt of the acid. Even after all of the phosphate groups have reacted with Ca and Mg, the phytate is still capable of chelating iron, zinc, and copper; its ability to chelate is enhanced in the presence of excess calcium. Phytate has been experimentally removed from soyabean meal or other vegetable-based ingredients by reaction with the enzyme phytase. Growth studies with rainbow trout using OMP-type diets in which 50 percent of the fishmeal was substituted with dephytinized soyabean meal are shown. Feeding untreated soyabean meal resulted in 25 percent reduction in growth when compared to the OMP control. Dephytinized soyabean meal gave growth values of about 8-10 percent better than untreated soyabean meal. Total substitution with untreated soyabean meal produced high mortality in the fish after about 90 days. When the zinc, iron, and copper contents of the blood of the dead fish were determined, it was found that Zn and Fe levels were significantly lower than normal. Furthermore, feeding purified (casein-based) diets containing 0.5 percent of either Ca or Na phytate to rainbow trout resulted in a 10 percent reduction in growth. No other obvious physiological defects were noted in the diets containing phytates. Compounds other than phytates present in soyabean meal are also capable of binding trace nutrients. Recent works by other investigators strongly suggest that diets containing proteins of vegetable origin will perform more satisfactorily if they are fortified with Zn and Fe. Recent laboratory data show, for example, that heating soyabean meal also reduces its iron-binding ability so that the improvement in performance of heated soya is not only related to inactivation of trypsin inhibitors and the denaturing of protein but also to the increased bio-availability of micro-nutrients. The problems of soyabean meal, particularly those concerning phytates, are associated more or less with all cereal-based components. The absence of phytase in the soyabean, however, may make the problem with soyabean meal more severe than with other cereal-based feedstuffs such as wheat.

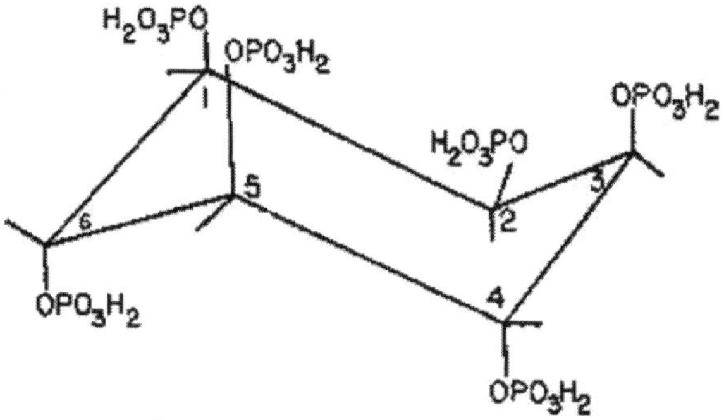

Figure 1: Structure and configuration of phytic acid (Myoinositol)

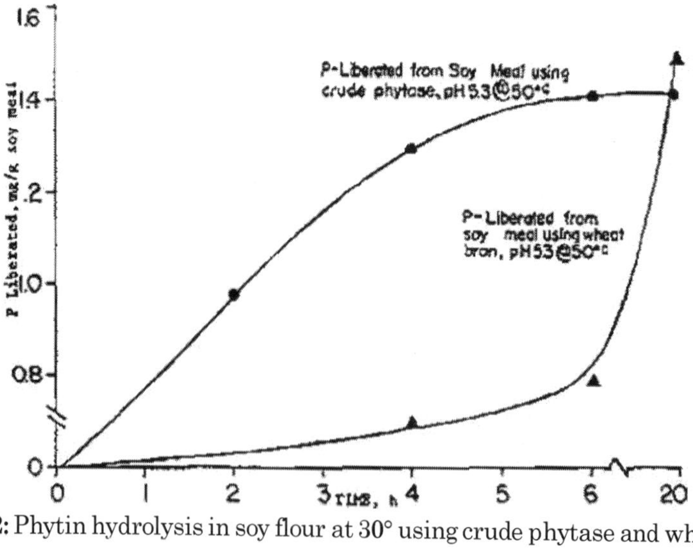

Figure 2: Phytin hydrolysis in soy flour at 30° using crude phytase and wheat bran

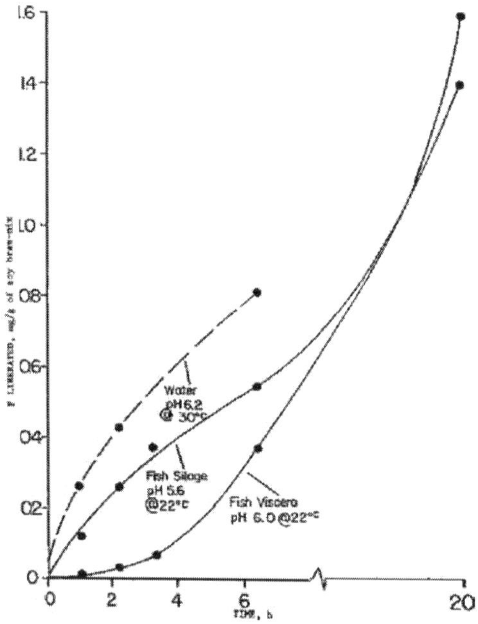

Figure 3: Phytin hydrolysis in soy flour using wheat bran as phytase source

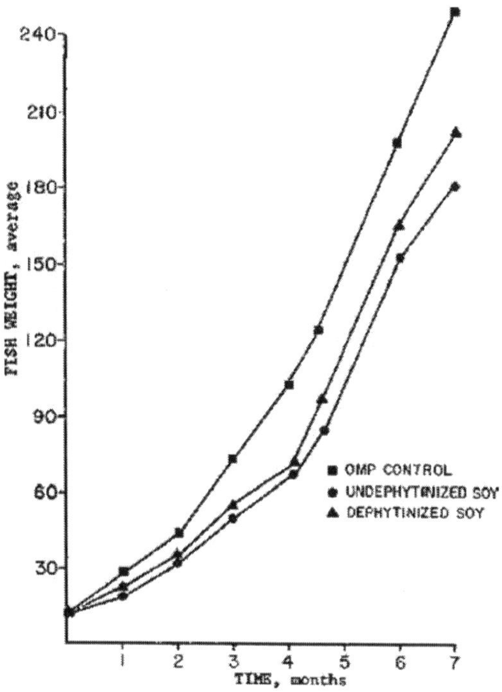

Figure 4: Growth of rainbow trout fed OMP and OMP-type diets in which 50% of the fish meal was substituted with undephytinized and dephytinized soy

Table 1: Calcium, Copper, Iron, and Zinc Content in the Blood of Rainbow Trout Fed OMP and OMP-Type Diets in which 100° of the Fish Meal Portion of the Diet was Substituted with Soyabean Meal and Dephytinized Soyabean Meal

Diet	*Element, ppm*			
	Calcium	*Copper*	*Iron*	*Zinc*
Control (OMP)	71 (71-72.)	1.5 (1.3-1.4)	213 (204-222)	14.8 (14.5-15.2)
OMP - no fish meal 100% soyabean meal	73 (52-94)	22.2 (1.5-2.7)	141 (92-220)	9.1 (5.6-12.8)
OMP - no fish meal 100% dephytinized soyabean meal	66 (43-76)	1.7 (1.2-2.2)	190; (157-240)	10.1 (9.8-11.9)

Single Cell Protein

Predictions of future protein shortages have spurred research on non-agricultural methods of protein production. As a result, certain types of high protein products suitable for feeding to livestock and fish are now produced on an industrial scale. Notable among these new protein sources, are single-cell organisms. Man has used single-cell organisms in connexion with his food supply for centuries, primarily in fermentation processes. The direct consumption by man and animals of single-cell organisms, however, is a recent innovation. More commonly referred to as single-cell protein (SCP) because they are high in protein contents this class of feeds tuffs are mainly derived from unicellular organisms such as yeast and bacteria, but can also include fungi and algae. SCP have reasonably well balanced amino acid profiles. When added to the diet of trout, some produced growth comparable to fish meal. SCP are an excellent source of some vitamins and minerals and also possess usable lipids and carbohydrates.

Two commercial processes have been developed for large-scale production of SCP. One uses n-paraffin (C-10:C-23 range) as a substrate for growing yeast. The yeast organism most commonly cultured is Candida lipolytica. The other process uses methanol as a substrate and the organism of preference is Methylophilus methylotrophus. Both processes are similar in nature and the process for producing SCP of bacterial origin is schematized. To date, it appears that the process for SCP of bacterial origin is preferred for two very important reasons:

(i) Bacterial SCP produced from methanol is more easily purified. Yeast plants in-Italy and Spain have been closed down because of substrate contamination of the product.

(ii) The bacteria process operates at higher temperatures, 40-50° vs 37° for the yeast process. Since refrigeration is a major cost in operating the plants, operating temperature becomes a vital economic factor in SCP production. Bacterial SCP also has a higher protein content, about 70 percent as compared to 60 percent for yeast.

Table 2: Feed Conversion of Rainbow Trout Fed Experimental Diets Containing Single Cell Protein (Bacterial Origin) for 155 Days

	Feed Conversion	
% Substitution SCP	*SCP Added to Diet*	*SCP Replacing Diet*
0	1.22	1.24
25		1.25
50	1.26	1.26
75	1.26	1.32
100	1.26	1.29

SCP has been extensively evaluated with animals such as pigs, poultry, and calves; excellent results have been reported. Tests with fish have been conducted primarily with salmonids with variable results. In general, yeast SCP can be substituted for about 25-40 percent of, the fishmeal in standard OMP-type diets.

To date, inconclusive results have been reported with SCP of bacterial origin. Successful substitutions have been made of the entire fish meal component with SCP bacterial in OMP-type diets for rainbow trout in fresh water. When these same diets were fed to coho salmon in salt water, it was possible to substitute only 25 percent of the fishmeal without adverse effects on growth and feed conversion. Fortification of the diets with iron, phenylalanine, and/or methionine did not alter growth or feed conversion. At present, tests are in progress to determine whether the dietary requirements for coho are different from rainbow trout or whether the salt water environment is a factor in the tests. One possible factor relating to the poorer performance of the bacterial SCP is the Ca/Mg ratio of the diet. OMP control diet contains a Ca/Mg ratio of about 6:1. Analyses of the bacterial SCP diets (100, percent substitution) show them to contain a Ca/Mg ratio of 16:1.

In discussing the use of SCP, one should not lose sight of the fact that spent brewer's yeast (BY) is a SCP source and also an article of commerce. It is often used in small quantities in preparing trout rations. Brewer's yeast contains about 45 percent protein, 8 percent fat, 13 percent ash, 10 percent water, and about 23 percent fibre and carbohydrates. It has an excellent amino acid profile, being deficient only in methionine. Recent work at the University of Colorado, however, has shown that when brewer's yeast contributed over 15 percent of the protein in the diet, growth of fish began to decrease. At the 50 percent level, growth was reduced by 30 percent as compared to fishmeal. These investigators showed that the primary problem was that of digestibility, the BY being only 53 percent digestible, with only 76 percent of the protein being digestible. Here again, the problem of the digestibility of a feed ingredient is an important factor in assessing the value of alternative feed sources.

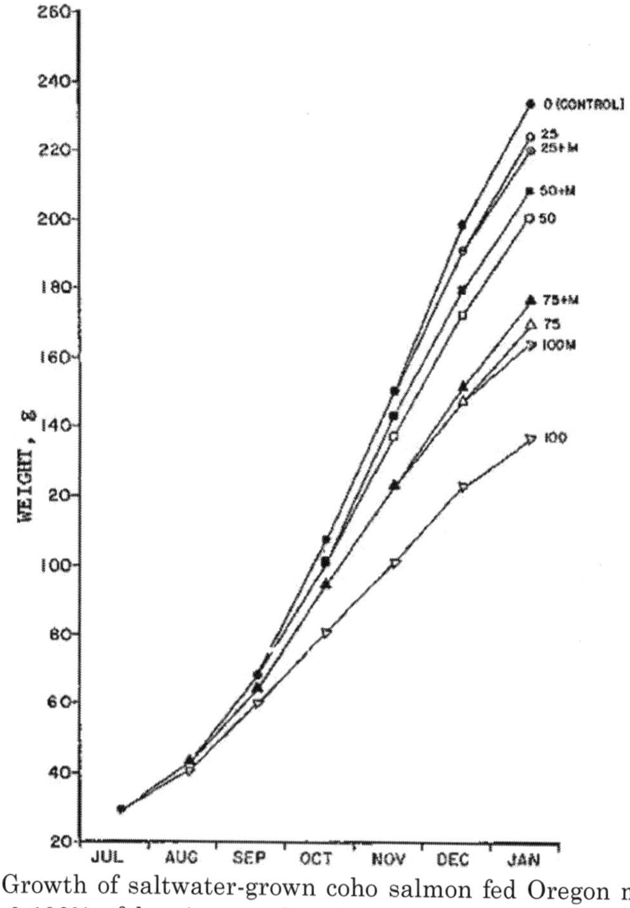

Figure 5: Growth of saltwater-grown coho salmon fed Oregon moist pellet diets with 0-100% of herring meal replaced with yeast SCP. M indicates methionine fortification

SCP in the form of dried sludge has been evaluated. The sludge consisting of microorganisms grown on paper mill waste had a composition of 43 percent protein, 0.43 percent fat, 3.0 percent ash, 28 percent fibre and 15 percent carbohydrate. When substituted at the 25 percent level in trout diets containing 32 percent herring meal, growth and feed efficiency were significantly lowered; i.e., feed conversion went from 1.45 to 1.58 while percent daily weight gains fell from 18,2 to 13.7 percent, respectively. Amino acid analysis of the SCP used in these tests showed it to be deficient in phenylalanine, methionine, and arginine. SCP grown on potato wastes, on the other hand, could be substituted for about 40 percent of the total protein of OMP diets without significant loss of performance. A danger associated with the use of SCP grown on industrial wastes is the possible risk of contamination with mycotoxins - aflatoxin, of course, being the more serious.

Krill

Of all the unconventional feedstuffs of animal origin, krill represents the largest potentially available source. Various estimates made by fishery biologists show that perhaps over 300 million metric tons could be harvested annually. The problem here, of course, is that krill is widely dispersed, primarily in the Arctic and Antarctic regions, and would require a tremendous outlay in capital and energy for the harvest. Recently, however, there has been a renewed interest in krill fishery with the principal investigations being carried out by the Federal Republic of Germany, Japan, Poland, and the U.S.S.R. Krill meal has been prepared and its reported composition is 55 percent protein, 10 percent moisture, 10-15 percent fat and 15.2 percent ash. The remaining portion of the composition is probably chitin. Analysis of Russian and Polish krill show krill to contain about 1.5 percent chitin on an as-received basis. Krill has been fed in diets to rainbow trout. A diet containing 35 percent starch and 7.0 percent oil was almost (within 5 percent) equal to a control diet containing 70 percent fishmeal, 24.4 percent starch and 4.1 percent oil. When krill was the sole source of protein, growth of the fish was reported to be lower than the fishmeal-fed control.

When isocaloric diets in which the protein contents were solely derived from krill or fishmeal were fed to channel catfish for 19.0 weeks, growth on the krill-fed fish was about 80 percent of those fed the fishmeal diet. Analysis of the amino acid contents of krill meal and zooplankton such as red crab (Pleuroncodes planipes and Euphausid pacifica) show that amino acid profiles of these zooplankton approximate those of herring meal. A possible explanation for the inferior performance of krill is probably related to its high ash content. A process for reducing ash content of crustacea meals by a milling and screening technique has been reported that not only reduces the ash but also increases the protein level of the final product.

High ash content, particularly calcium, has been shown to reduce the bio-availability of manganese and zinc in animal feeding studies. If krill could be economically harvested and processed at sea the Ca content could be reduced by a procedure developed for the processing of red crab. In this process, the chitinous fraction is reduced by extracting the krill through a flesh-separating machine, or any other appropriate device such as a Beehive separator. The objective in developing this process which yields several products (i.e., meal, chitosan and protein) was to improve the economic prospects of krill utilisation.

In feeding salmonids, krill provides an important source of carotenoids. Salmonids are incapable of synthesizing carotenoids and their characteristic red colour can only be derived from ingested carotenoids. In both Europe and the USA, carotenoids are formulated into the diets by the addition of crustacea and/or synthetic carotenoids such as canthaxanthine.

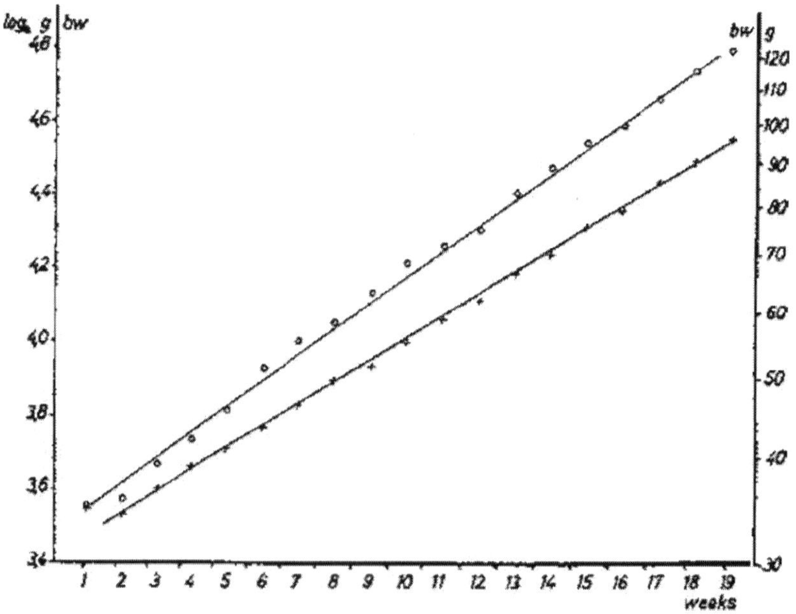

Figure 6: Growth comparison of channel catfish fingerlings fed with a fish meal diet. (0) and krill meal (+), average weight weekly. (Adapted from Hilge, 1978)

Although canthaxanthine offers ease of formulation, about three times more is needed in the diet to impart the same degree of pigmentation as compared to astaxanthine and astaxanthine esters. It also has a stronger tendency to fade during cooking.

Carotenoids are labile compounds and are prone to degradation by heat, acids, alkalis, and oxidation. They can be somewhat stabilised by antioxidants such as ethoxyquin. A process for producing a stabilised carotenoid extract was developed and is shown.

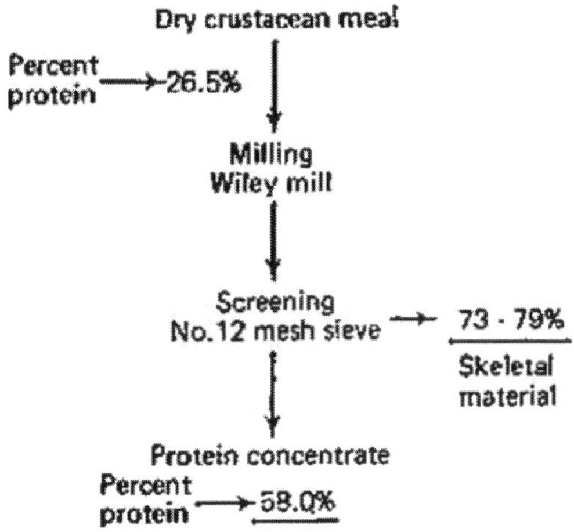

Figure 7: Decalcification of crustacean meals

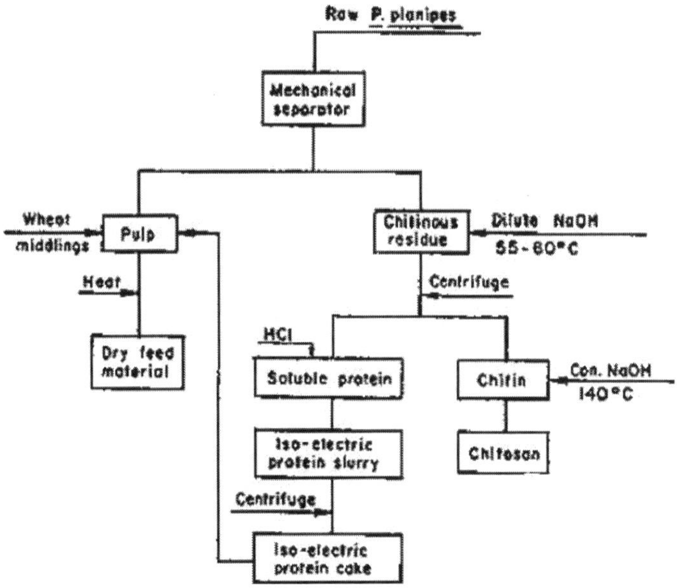

Figure 8: Processing of red crab (P. planipes) into a feed material, chitin, and/ or chitosan

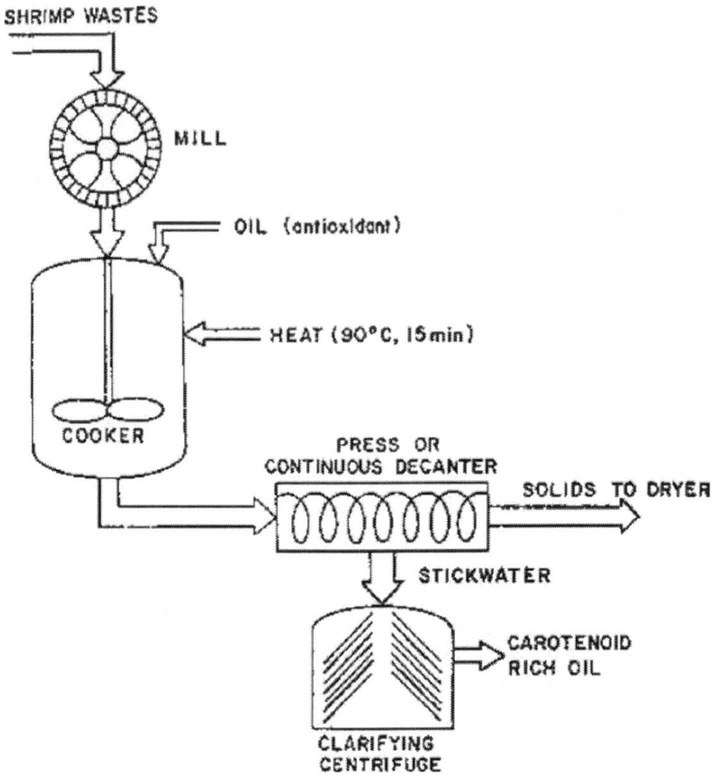

Figure 9: Preparation of carotenoid concentrates by extraction of shrimp waste with soy oil (Northwest and Alaska Fisheries Centre)

Poultry By-Products and Feather Meal

Poultry by-products and feather meal have long been articles of commerce and their methods of production have been discussed elsewhere. Poultry by-products are primarily used by the pet food industry, and feather meal is a dietary ingredient, in poultry rations. They appear to be excellent protein and lipid sources containing 69 percent crude protein, 10-21 percent lipid and about 10 percent ash. The so-called hydrolysed feather meal is really a misnomer. It is at best a slightly hydrolysed product produced by cooking feathers in the presence of calcium hydroxide to increase its digestibility. Little, if any, free amino acids are found in this product. Feather meal has been evaluated in both fresh and salt water species. It contains about 80-85 percent protein and is a relatively good source of sulphur-containing amino acids. Whether these amino acids are completely available has not been demonstrated. Good results have been reported when it is used in catfish diets at the 15 percent level. Poultry by-products are lower in lysine than fishmeal, and trout diets containing over 75 percent poultry by-products would be deficient in the amino acid.

That some of the sulphur-containing amino acids in hydrolysed feather meal are available to salmonids was recently demonstrated in feeding tests at the University of Washington. In these experiements, trout grew much better on mixes of feather meal and yeast SCP than when either of these commodities was used as the sole source of protein.

Other Potential Feedstuffs

Other feeds tuffs that may be potential dietary ingredients for fish include rapeseed meal and sunflower meal. These commodities contain from 37 to 43 percent protein. Rape-seed meal has been evaluated recently. Inclusion of up to 22 percent in a basic diet containing 40 percent fishmeal produced satisfactory results in the coho salmon. Because rapeseed meal replaced brewer's yeast, cottonseed meal, and wheat shorts in the test diet, the results were no indication of its true performance as compared to fishmeal. The same study also showered that, although the amino acid profile of rapeseed meal was similar to soya-bean meal, its higher fibre, tannin, and phytic acid contents could pose a distinct deterrent to its use in aquaculture diets. The high glucoinoleate content in rapeseed could also give rise to isothiocyanates and gastrin formation in the gut thereby interfering with thyroid function. Feeding tests indeed showed that, although the diet containing 22 percent rapeseed meal did not markedly decrease growth and feed efficiency, it did result in a compensatory increase in thyroid activity.

A new approach to a rather old process was recently tried; i.e., it was thought that perhaps insect larvae, if properly cultivated and formulated, could serve as an alternate source of nutrients in aquaculture diets. House fly (Musca domestica) larvae were cultivated and tests were made to evaluate their use in OMP-type salmonid and tilapia diets.

The larvae were grown by incubating fly eggs on a substrate composed of two parts wheat middlings, one part alfalfa, and three parts water. The fly egg-substrate mixture was placed in partially covered aluminium containers to a depth of 15 cm and incubated at 30°C for 4-5 days.- After incubation, the substrate and larvae were dried in a rotary vacuum dryer (735 mm of vacuum, 85-100°C) to a moisture content of approximately 8 percent. After drying, the larvae were separated from the substrate by screening. The mean proximate composition of the dried fly larvae was:

moisture	8%
protein	45%
fat	15%
ash	8%
chitin	25%

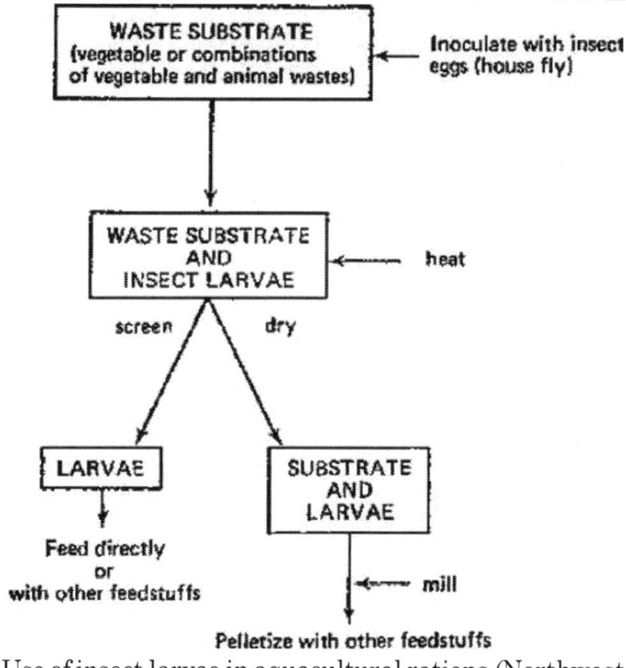

Figure 10: Use of insect larvae in aquacultural rations (Northwest and Alaska Fisheries Centre)

Amino acid composition was about the same as fishmeal. Prior to formulation into diets, the larvae were milled by passing them through a hammer mill equipped with a 16 mesh (1 mm) screen. Isonitrogenous diets were prepared by replacing appropriate quantities of fishmeal in the control diets with purified chitin. These diets were tested on coho salmon and rainbow trout.

Coho salmon were fed a ration ranging from 1 percent to 2 percent of body weight (depending on water temperature) per day. The ration was adjusted every seven days to produce, as closely as possible, the optimum feed conversion. Rainbow trout were fed a ration of 2.5 percent of body weight per day. Ration adjustment was made every seven days.

Rainbow trout fed diets containing 25 percent, 50 percent, and 100 percent substitution of fly larvae for fishmeal produced the same growth and feed conversion were determined on two sets of diets (i.e., those in which fly larvae was substituted on an equivalent weight basis and another in which the OMP control diet was prepared on an isonitrogenous basis by the substitution of fishmeal with purified chitin. In all experiments, growth and feed conversion (calculated on an isocaloric basis vs direct measurement) were identical.

In another project, diets were prepared by drying the larvae and substrate without separation. After drying, the mixture was milled and directly pelletized without the use of any binders. The pellets produced were very hard and even tilapia did not readily eat the pellets. The project was not pursued further, but it does show promise as a method for upgrading various wastes for use in aquaculture rations. The possibility of using the larvae directly for larval and immature species, such as shellfish species, also presents interesting possibilities. For example Macrobrachia found fly larvae very much to their liking.

Table 3: Comparative Amino Acid Profile of the Proteins of Fishmeal and Fly Larvae

Amino acid	***% Protein***	
	Fishmeal	***Fly larvae***
Alanine	6.34	6.15
Arginine	5.82	5.42
Aspartic	9.35	10.8
Cystine	0.70	0.82
Glutamic	13.3	12.2
Glycine	5.90	5.40
Histidine	2.22	3.50
Isoleucine	4.85	4.13
Leucine	7.35	6.95
Lysine	7.85	7.37
Methionine	2.84	2.24

Phenylalanine	4.35	6.95
Proline	4.35	3.66
Serine	4.55	4.51
Threonine	4.55	4.53
Tryptophan	1.33	1.45
Tyrosine	3.45	8.10
Valine	5.65	5.60

Table 4: Feed Conversion of OMP and OMP-Type Diets Containing Fly Larvae Fed to Rainbow Trout for 155 days

Diet fed	*Feed conversion*
OMP control	1.23
25% fly larvae	1.29 (1.24)
50% fly larvae	1.36 (1.23)
100% fly larvae	1.55 (1.26)

Values in parenthesis represent feed conversion at iso-caloric basis

It should be pointed out again that the production of mould or bacterial toxins prohibit this type of approach.

Waste Recycling and Direct Fertilization

The potential of producing natural feeds for aquaculture by direct fertilization of the aqueous environment or by recycling wastes is described in the following publications: (a) Problems and Potential of Recycling Wastes for Aquaculture, by J. Kildow, Massachusetts Institute of Technology, Massachusetts, 1974, (b) Effccts of Feeding, Fertilization and Vegetation in the Production of Red Swamp Crayfish, by F. dark, J. Avault and S.P. Meyer, Department of Food Science, Louisiana State University, Baton Rouge, Louisiana, 1974, (c) Preliminary Results with a Pilot Plant Waste Recycling Marine Aquaculture Systems, by J. Ryther, Woods Hole Oceanographic Institute, Woods Hole, Massachusetts, 1975.

The above works relate to processes that are designed to supply surplus nutrients through fertilization, either directly or indirectly via waste recycling. Theoretically, these processes are most efficient if they are applied to polyculture systems. In actual commercial practice, however, a combination of prepared foods and naturally grown feed may provide the best promise for this system. Experiments conducted in Louisiana, USA, showed that optimum yield with crayfish was obtained when ponds were fertilized with N-P-K (8:8:8) at 130 kg/ha in combination with pelleted feeds. The fertilizer was applied in November, in March, and again in May and June. The vegetation in the pond consisted of phytoplankton, algae, weeds, and smart weed. Fertilizing the pond gave a 13-27 percent higher yield than without fertilizer. Whether the crayfish consumed phytoplankton or fixed plant growth did not seem to matter since other zooplankton consumed the plants which, in turn, were eaten by the crayfish. Both Kildow (1974) and Ryther (1975) presented similar systems. Basically these systems are depicted. Actually, these processes were developed to produce a biological tertiary sewage treatment process capable of removing all inorganic nutrients (primarily N and P) from secondary effluents prior to discharge into the environment (in this instance, the sea). The secondary objective was to develop a marine aquaculture system of a primary crop of shellfish and a secondary crop of other commercially valuable marine organisms such as finfish or shellfish. This, of course, would help defray the cost of the tertiary treatment process.

Treated sewage, therefore, is used to grow phytoplankton. The shellfish then removes the phytoplankton from the water. Solid wastes produced by shellfish and uneaten phytoplankton support dense populations of small invertebrates (amphipods, worms, etc.). These can then serve as food for a secondary crop of lobster or flounder.

The problems related to utilising wastes, however, are numerous; some of these are:

(a) health hazards

(b) aesthetics

(c) environmental impact

Some marketing studies in the USA have already indicated that it would be difficult to market fish grown under conditions described here. As for the reality of toxin, metal absorption or accumulation, work done at this laboratory has shown that toxic metals such as mercury are rapidly deposited in the flesh and kidney of salmonids when diets containing the metal are fed. In experiments using dogfish meal as the mercury source, it was shown that this accumulation occurred very rapidly in the fish consuming diets at all replacement levels of fish meal by dogfish meal.

With respect to other toxic compounds, work in the UK has demonstrated that effluents of pea silage could be extremely toxic to fish (carp and goldfish). Several types of toxic compounds have been found in these effluents; e.g., volatile fatty acids, sulphites, phenols, and ammonia. Strangely enough, the most toxic factor was the fatty acids which it is assumed were probably found in the greatest amounts. There was also a high percentage of p-cresol which might have been formed by the action of bacteria on tyrosine. Although the purpose of the UK study was to determine the possible toxic effects of pea silage effluents that might find their way into streams, this study would strongly suggest that if direct fertilization of ponds is planned, care should be exercised as the risk of introducing toxic substances via the wastes can be great.

Factors Affecting the Organoleptic Properties of Fish

Diet does influence the organoleptic quality of cultured fish. The factors that contribute to the organoleptic quality of fish are as numerous and complex as those associated with other foods. Unlike other foods, however, little attempt has been made to control these organoleptic factors. With wild fish, flavour and general overall acceptance have been associated with its texture, flesh composition (fat and protein contents), and the characteristic flavour constituents that are associated with the respective species. In marine species, it has long been recognised that their flavour is influenced by seasonal and environmental influences. Environmental factors include not only dissolved substances and plant nutrient but also the available animal food supply (food chain), the abundance of which is influenced by such factors as water temperature, water cycle, and photo period. Water temperature influences the fatty acid composition in muscle tissue of teleost fishes. At lower water temperature the fish tend to accumulate more unsaturated fatty acids in their tissues and organs. The relation of fatty acid composition to flavour and textural alteration during processing and storage, of course, is well recognised. It becomes quickly apparent, therefore, that changes in environmental factors need to be minimised when attempts are made to control the organoleptic qualities of cultured fish via the diet.

A broad prospective of the environment as it relates to probable influences on the organoleptic properties of cultured fish is depicted. It should be noted that two environments must be considered: the total environment, consisting of oceans, lakes, rivers, ponds, etc., and the local environment (i.e., the location of the fish with respect to the total environment). The type of fish culture, of course, will determine the degree that these environments influence the organoleptic properties of the fish. A review of the literature reveals that most of the research on the flavour attributes of cultured fish is related to environmental influences rather than specific nutrients. This is particularly true of pond-cultured species where, for obvious reasons, fish are reared in near static water conditions. In this type of environment, the growth of flavour-producing organisms, decomposing plant growth, and unconsumed feed all contribute to the organoleptic characteristic of the fish.

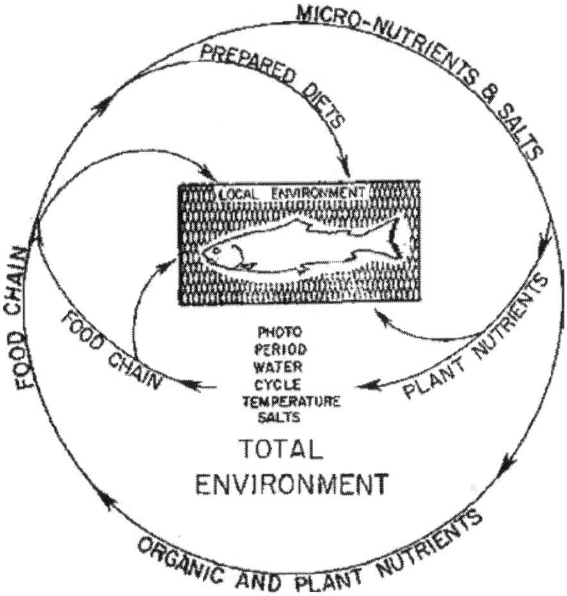

Figure 11: Environmental and dietary influences affecting the organoleptic properties of cultured fish

The organoleptic characteristics of European carp grown in ponds fertilized by liquid manure and those grown on diets of grain or high protein pellets have been compared. Carp grown in the fertilized tanks were claimed to be superior in flavour and colour to those fed the prepared diets. However, fish grown in fertilized ponds had a flesh fat content of about 6 percent, while those fed prepared diets ranged from 14 percent to 22 percent fat. It would appear that, all other factors being equal, the compositional constituents in the flesh more strongly influenced the eating quality of the fish than did the flavour components contributed by the diets.

In contrast to pond-reared fish, marine species such as salmonids are generally grown in non-static water, thus decreasing the effects from environmental factors. But even in a marine environment, it must be kept in mind that environmental factors can easily override dietary factors when attempts are made to control the organoleptic qualities by dietary means. Fish have been described by one aquaculturist as biological sponges that can absorb many inorganic and organic materials through their gills as well as their intestinal tract. Indeed, catfish have been shown to absorb 2-pentanone and dimethylsulphide within 10 minutes after being exposed to their solutions. Hence, while rigid quality control measures are needed to. prevent off-flavours, desirable flavours could be induced by exposing live fish to potentially beneficial solutions. A recent review of environmental and seasonal factors that influence the flavour attributes of the cod claimed that these qualities are related not only to the amount and type of feed consumed, but also to the composition of the lipid that, in turn, is influenced by feed and sexual maturity of the fish. An interesting point brought out by the author is that the post-mortem pH of the muscle is influenced by food supply, lipid content, and swimming activity of the fish.

Work covered in the literature definitely shows that consumers and taste panelists can distinguish between wild and cultured fish. Ignoring "off" or foreign flavours, however, it has not been established whether the consumer prefers wild fish, bearing in mind that this is true only when high quality cultured fish of the same size are compared against the wild fish. -In this respect, it had been shown that wild trout possessed distinctly different organoleptic properties from hatchery trout fed a diet consisting of beef, pork spleens, and horse meat. While the wild fish were consistently rated superior in flavour and aroma and texture to the hatchery trout, there was also a considerable difference in the organoleptic factors of the hatchery fish as they were taken from different hatcheries. Again, this indicates environmental conditions playing a major role in governing organoleptic qualities of fish.

In conclusion, it is believed that one can safely say that organoleptic changes, whether they be due to diet or environment, are readily detectable, but whether these changes affect acceptability of the fish has yet to be assessed. No doubt efforts need to be made to maintain some of the basic organoleptic characteristics, such as colour and texture traditionally associated with wild species. One for example, would not question that salmon possessing the characteristic red colouration would have a much wider acceptance than those having no colour. Similarly, but not as pronounced, we would expect that a firm texture in finfish such as salmonids or carp would enjoy a wider degree of acceptance than those possessing a mushy texture. The effect of different flavour notes, except when they are distinctly foreign to the fish, however, are as difficult to assess as they are to control. The seemingly simple expedience of controlling the fat content may prove to be difficult. Some work already shows that changing the lipid content in a salmonid diet does not always significantly or uniformly change the lipid content in the salmon. It would appear that water temperature, growth rate, stocking density, and sexual maturity are all involved in controlling this variable.

Practically all the results in the literature today show that the quality of cultured fish with respect to wild fish should be improved. It is unlikely, however, that this quality improvement will be achieved with the feeds that are in current use. These feeds, after all, were developed to provide optimum growth and feed conversion in the fish. As more is learned about fish nutrition and feeding, it may turn out that supplementary dietary rations will have to be developed that have as their specific function the ability to favourably alter the organoleptic characteristics of the cultured fish. In this area, it is expected that a primary use for alternate or unconventional feedstuffs -for cultured fish will be developed.

Storage Problems of Feedstuffs

Losses occurring in feedstuffs during storage fall under four major categories:

(a) weight loss,

(b) quality loss,

(c) health risk, and

(d) economic loss.

These losses arise from the foraging activities of insects, micro-organisms and animals; improper handling; and, physical and chemical changes, all of which are interrelated. Storage loss in a feed mill is primarily due to material eaten or destroyed by insect and animal pests and fungi. When serious infestation by these pests occurs there is extensive weight loss accompanied by damage to quality. Intense insect activity often results in mould growth which not only completes the destruction of the feedstuffs but also poses serious health risks to animals or fish feeding on rations containing damaged feed ingredients.

The first three categories of storage loss are of primary concern to the compound feed manufacturer. Weight loss due to loss of moisture content or to the presence of a large insect population in stored feed is problematic in developing countries where post-harvest handling and processing are often improperly conducted. Ineffective enforcement of quality standards (if, in fact, such standards exist) results in the production and supply of inadequately processed feed ingredients which often are more susceptible to deteriorative processes. The lack of quality standards reflects the relative unimportance attached to aesthetic considerations in commercial transactions of feed commodities in these countries. Hence, economic loss which we shall define here as the cost incurred in inspection, prevention, and control to maintain quality standards (although considered to be of overriding importance in most industrialised countries) is superseded by the other three major categories of storage losses in developing countries.

This chapter looks at the effects of insects and micro-organisms and the deteriorative changes that result from their presence in stored ingredients in feed mills in the tropics.

Insects

Insects feed on most feed ingredients and contaminate them with faeces, webbing, body parts, foul odours, and micro-organisms. Beetles and moths are the most destructive of the grain insects, and many are capable of destroying an entire store of feed.

Factors Affecting Insect Infestation of Feedstuffs

The occurrence and development of an insect infestation is dependent on many factors such as source of insects, available food, temperature, moisture, air, condition of the feed-stuff, presence of other organisms, and the efforts to exclude or kill the pests. Major factors affecting population growth of most insect species are: temperature, relative humidity, and moisture content of the feed ingredient. The nutritive content and certain physical properties of feeds will also determine the vulnerability of such materials to attack.

Only a few insect species are able to attack sound kernels of feed grains. High moisture content (16 percent or more) renders feed grains soft and susceptible to attack. Meals pressed into cakes or hard flakes are more resistant. Insects appear to eat small particles more readily than large ones.

Table 5: Environmental Constraints of Major Insect Pests Infesting Feedstuffs

Species		*Susceptible feedstuffs*	*Minimum for increase to epidemic numbers*		*Optimum range for increase*
Scientific name	*Common name*		*Temp, °C*	*Rel. Hum, %*	*Temp, °C*
Sitophilus spp.	Weevils	Cereal grains	15	50	26-30
Sitotroga cerealella	Grain moth	Cereal grains	16	30	26-30
Tribolium spp.	Flour beetles	Cereal grains and cereal by-products, oilseed cakes and meals, finished feed	21	*	30-33
Oryzaephilus spp.	Saw-tooth grain beetles	Cereal grains and cereal by-products, oilseed cakes and meals, finished feed	21	10	31-34
Cryptolestes spp.	Flat grain beetles	Cereal grains and cereal by-products, oilseed cakes and meals, finished feed	21	50	30-33
Cadra cantella	Tropical warehouse moth	Cereal grains and cereal by-products, oilseed cakes and meals, finished feed	17	25	28-32
Rhizopertha dominica	Lesser grain borer	Cereal grains, pulses, dried roots	23	30	32-35
Trogoderma granarium	Khapra beetle	Cereal grains and cereal by-products, oilseed cakes and meals, finished feed, pulses	24	*	33-37

* Species breeding rapidly even in driest conditions

(Sources; Howe, 1965; Cockerel 1 et al, 1971)

Temperature

All insects attacking stored feedstuffs have an optimum zone of temperature at which populations increase most rapidly. Most of the important insect pests are tropical species with an optimum temperature of about 28°C. It is, therefore, evident that losses from insect infestation will be greatest in the tropics.

Relative Humidity

As a rule, relative humidity affects the rate of population increase in insects less dramatically. Up to 70 percent relative humidity, there is progressive increase in insect multiplication. Beyond 70 percent relative humidity, mould formation sets in and complicates the situation. The moisture content of feedstuff is closely related to the relative humidity. A low moisture content coupled with low humidity will provide protection against insect infestation.

Feeding Habits of Insects

Some insects are not discriminatory in their feeding habits whereas others are highly selective in their choice of the parts of a feed ingredient eaten. For example, moth larvae generally feed on or near the surface of the grain mass; beetles are destructive throughout. Most larvae spread silk webbing over the surface of feed grains and their by-products, making them not only unsightly but also creating handling and processing problems as well.

Insect infestations sometimes cause excessive heating of grain. When the insect population reaches a certain density, their metabolic activities release more heat than can be dissipated. In localised areas where the insect population is extremely dense the temperature may reach 45°C. Associated micro-organisms, mainly fungi, may raise this to nearly 75 °C, causing extensive spoilage and, occasionally, spontaneous combustion.

Losses Due to Insect Attack

The best warning of serious weight loss is the presence of a large insect population. This is easily identifiable in sacked grain by the appearance on the sack surface of frass resulting from feeding activity of the insects. High temperature, high humidity, softness, and high nutritive value of the feedstuff and storage in small quantities are all conducive to insect damage, but are often unavoidable. Loss may be aggravated by prolonged storage. Failure to keep the storage area clean and retention of infested sweepings will increase the liability of insect attack.

Weight Loss

Weight loss in infested feed is not always evident unless it involves sacked grain or oil cakes when the appearance of frass on the surface of the sack points to the feeding activity of a large insect population.

Quality Loss

For the feed manufacturer, one of the most important considerations of insect attack on feedstuffs is the loss of quality. The effects on quality are various. Most stored feed material undergo some chemical changes that alter their flavour and nutritive value. As well as eating some of the feed, insect pests tend to accelerate these harmful chemical changes. Secretion of enzyme lipase by the insects themselves will enhance deteriorative chemical processes. Few stored feedstuffs are homogeneous and insects and mites are known to select those parts or particles of feed they prefer.

Many feedstuffs contain a high percentage of fat which tends to break down during storage. This breakdown is accelerated by insect attack, especially when the insects break off small particles, introduce micro-organisms, or raise the temperature or moisture content. Evidently, the insects use the fat in the material they eat. The breakdown of fat causes an increase in free fatty acids which cause off-flavours. The free fatty acid involved in product rancidity is assumed to be oleic acid, and the quantities released is a result of oxidation of fats in certain feedstuffs are quite substantial.

The subject of lipid oxidation will be further discussed when the deteriorative changes taking place in stored feedstuffs are considered.

Scavenging insects, such as cockroaches, may cause contamination with pathogenic bacteria such as Salmonella.

Control

Climate is the most important factor determining the effectiveness of a storage system for feed ingredients because of the close relationship between insect growth and ambient climatic conditions. An important relationship also exists between insects and micro-organisms in stored feedstuffs.

Total eradication of insect populations from tropical warehouses for feed ingredients is not possible. The degree of infestation can, however, be brought to manageable proportions through a programme of vigilance and effective control measures. Heavily infested ingredients should not be brought into the store. Infested material, if accepted, should be kept separate until fumigated (this is to be done as soon as possible) to totally eradicate the pests.

Good house-keeping involving the sweeping up of spilled material will check scavenger insects as well as rodents which may carry insects in their fur.

Micro-organisms

Micro-organisms are biological contaminants of the natural environment and are present in all feedstuffs. They persist after crops have been harvested from the fields and in animal carcasses prior to rendering. Because bacterial and field fungi do not thrive at moisture levels below 20 percent, post-harvest processing of commodities and animal renderings involving heat, chemical and mechanical extraction, and dehydration eliminate most of the original contaminating microflora. Fungi spores, which are resistant to harsh processing treatment, may remain dormant in the processed feedstuff until more favourable conditions once again permit their proliferation.

Factors Affecting Fungal Growth in Feedstuffs

Recontamination of feedstuffs by adventitious micro-organisms during storage is of primary concern to the feed processor. Adventitious storage fungi grow at moisture content (15 to 20 percent) in equilibrium with a relative humidity of 70 to 90 percent and are considered the principal spoilers of feedstuffs in storage. When the relative humidity falls below 65 percent no growth occurs.

Under favourable conditions, fungi can raise the temperature in their immediate environment to 55 C with con commit tent increase in moisture content of the affected feedstuff to as high as 20 percent. When this occurs, secondary spoilage by bacteria takes place.

The most common fungi involved in the spoilage of feedstuffs belong to the Aspergillus spp. and the Penicillium spp. These grow at temperatures up to 55 C and at a minimum, relative humidity of 65 percent. They are most destructive when temperatures exceed 25 C and relative humidity exceeds 85 percent.

Detrimental Effects of Storage Fungi on Feedstuffs

The chief effects of storage fungi on feedstuffs are:

(a) mycotoxin production,

(b) heating,

(c) moisture increase, and

(d) mustiness (staleness).

Mycotoxin Production

Mycotoxins are compounds produced by fungi growing in infested agricultural commodities. They are toxic to both humans and animals.

The aflatoxins, a group of highly toxic and carcinogenic metabolites produced by Aspergillus flavus are perhaps the most important among mycotoxins contaminating feedstuffs. Transmission of the toxin to milk, meat, and eggs through farm animals feeding on contaminated feed poses an increasing health hazard. Studies on the toxicity of aflatoxin to fish have not been extensive but toxicity to trout (oral LD_{50}: 0.5 mg/kg body weight) is comparable to toxicity to ducklings, the latter recognised as the most susceptible animal to aflatoxin poisoning.

Table 6: Oral Toxicity of Aflatoxin to Various Animal Species

Species	*Age (weight)*	*Sex*	*LD_{50} in mg/kg body weight*
Duckling	1 day	M	0.37-0.56
Trout	100 g	M-F	0.5
Sheep	-	-	2.0
Pig	Weanling	M-F	2.0
Mouse	-	-	9.0
Rat	1 day	M-F	1.0
	21 days	M	5.5
	21 days	F	7.4

Feedstuffs known to be contaminated by A. flavus include: groundnut cakes, maize, sorghum, sunflower, cottonseed cakes, copra, and cassava. To produce aflatoxin, however, A. flavus must be present alone in a practically pure culture. The presence of other fungi, yeast, or bacteria seems to interfere with aflatoxin production. Crops such as peanuts, cottonseed, and copra are high aflatoxin risks precisely because A. flavus often infest them as a practically pure culture with few or no other microflora. In addition, the fungus produces the toxin in these crops at relatively low moisture levels, 9 to 10 percent, compared with 17 to 18 percent moisture for most feed grains. Feed grains such as maize and sorghum grown in the tropics, therefore, also pose high risk.

Heating and Moisture Increase

Mould growth in feedstuffs is accompanied by rising temperatures and moisture content. Aspergillus glaucus, which has a minimum moisture requirement of 14.5 percent, is the first significant species involved during mould infestation of feed grains. Temperature elevation that accompanies this initial attack flavours the proliferation of a second species, A. candidus, which raises the moisture level of the infested grain to 18 percent or higher. At such high moisture levels, A. flavus activity becomes intense and total destruction of the wholesomeness of the feed grain becomes complete.

Fungal activity in stored feed- grains is not often apparent until after serious damage is done. This is because such activity takes place not near the surface where temperature gradients produced by such activity are quickly abolished, but within the interior of the storage container. Silos for grains should, therefore, be equipped with temperature sensors to provide early warning of trouble. Similar preventive measures are not possible for bagged material. The common practice of storing bags of grains in large piles to minimise and control insect infestation actually promotes fungal activity, especially in the tropics. The "sweating" of bags within a large stack is evidence of serious fungal damage to the stored grain. The surface of such bags will feel warm to the touch, and when a hand is inserted into the grain, it will actually feel hot.

Staleness

Feedstuffs that are damaged by fungi tend to become lumpy. Feed grains suffer discolouration while damaged maize turns a dark brown with some blackened kernels being evident. The grain also exhibits a characteristic bluish sheen. Staleness or mustiness is another characteristic of commodities damaged by fungi.

Control

The prevention of mould contamination of stored feedstuffs depends mainly on the successful control of insect infestation, because the destructive activities of insects often create conditions favourable to mould; viz., increased moisture and temperature and the destruction of the protective hulls of feed grains expose their moist interiors. There is no effective way of eliminating mould, although effective measures have been developed to control their growth in compound feeds. These measures include the use of propionate and, more recently, gentian violet.

Deteriorative Changes in Stored Feedstuffs

Most stored feedstuffs undergo some chemical changes altering flavour and nutritive value. These changes are usually deteriorative in nature and are associated with the lipid content of the feedstuffs. Lipids tend to break down during storage into free fatty acids.

Factors Affecting Deteriorative Processes

Environmental Factors: Environmental factors that determine the extent of deterioration of stored feedstuffs also affect the rate of growth of its insect and microbial population. These include ambient temperature and relative humidity. Other environmental factors relate to the general cleanliness of the storage areas and to design features of the storage building; e.g., protection against rain and insulation against scavenger pests.

Insects and Micro-organisms

The role that insects and micro-organisms play in the breakdown of lipids in the feedstuffs they attack has been discussed earlier. The increase of fatty acids as a result of lipid breakdown is of particular importance in the storage of fish meal, cereal brans, and oil seed derivatives; e.g., copra, groundnut, and palm kernel meal. This increase in free fatty acids in improperly stored feedstuffs results in rancidity.

Rancidity

There are three major chemical processes that give rise to rancidity: oxidation, hydrolysis, and ketone formation. Due to the relative unimportance of the other two processes in stored feedstuffs only oxidation of fats will be described here.

Lipid Oxidation

Rancidity resulting from lipid oxidation is the most important deteriorative change occurring in stored feedstuffs. Feedstuffs containing lipids which are highly unsaturated (e.g., rice bran and fish meal), are especially susceptible to oxidation. The mechanism of lipid oxidation begins with auto-oxidation involving the direct reaction of lipids with molecular oxygen to form hydroperoxides. This is followed by secondary reactions yielding diperoxides if further oxidation takes place, or ketoglycerides if the hydroperoxides are dehydrated. Fission of hydroperoxides yield products containing carbonyl and hydroxy groups which will react further to form other products. Oxidation of carbon-carbon double bonds in other molecules gives rise to epoxides and hydroxy glycerides. These products of secondary oxidation of lipids contribute to 'off flavour' and include toxic compounds frequently-associated with rancidity. Furthermore, carbonyl groups produced by the fission of aldehydic hydroperoxides can react with the epsilon - amino group of lysine, thereby reducing the nutritive value of the protein.

Factors Affecting Lipid Oxidation

The chief factors increasing the rate of lipid oxidation in stored feedstuffs are as follows:

(a) *enzymes:* the presence of lipoxidase and perhaps other enzymes as well

(b) *hematin:* this factor is important in the storage of fish and meat meals

(c) *peroxides:* these compounds which are themselves products of auto-oxidation of lipids catalyse the oxidation of lipids

(d) *light:* ultra violet light is involved in the photolysis of peroxides

(e) *high temperature:* in general, the higher the storage temperature the more rapid the breakdown of lipids

(f) *trace metal catalysis:* many metals, notably iron, copper, cobalt, and zinc accelerate lipid oxidation. Iron and copper does this by direct electron transfer in redox reactions, while zinc induces the breakdown of hydroperoxide to free radicals.

Lipid oxidation can be inhibited by adding compounds known as anti-oxidants. Two commonly used feed anti-oxidants are ethoxyquin and butylated hydroxytoluene which sequester free radicals formed during oxidative processes. Cereal grains contain effective quantities of natural anti-oxidants (e.g. tocopherols) which impart considerable stability to their lipid contents, unless the kernels are damaged by storage pests.

Feedstuffs Especially Susceptible to Lipid Oxidation

Fish lipids are especially susceptible to oxidation due to the greater chain lengths of the fatty acids and the greater number of unsaturated carbon-carbon bonds along the fatty acid chains. Storage of fish meal is problematical, due to the frequent occurrence of exothermic oxidation which could lead to spontaneous combustion. The heat generated also leads to amino sugar reactions which lower the digestibility of protein. Most fish meals marketed commercially contain added anti-oxidant to inhibit lipid oxidation and thus reduce the risk of fire during sea transit and prolonged storage under unfavourable ambient conditions.

Other feedstuffs which have storage problems associated with high lipid content are rice bran and expeller oil cakes with high residual oil content. Among the latter, the most susceptible are copra cake and palm oil cake.

This is due to the significant levels of unsaturated fatty acids in coconut and palm oil. The lactose contained in milk powder has a tendency to react with loosely associated milk proteins to form melanoids in the Maillard reaction. These sugar-protein compounds have very low digestibility.

Storing Feedstuffs

Some weight loss of feedstuffs during storage is unavoidable. The extent of loss is affected by:

(1) the general hygiene of the store, because that determines whether or not insects can breed in the buildings away from the produce;

(2) the turn-over of the goods, because they determine the length of storage;

(3) the way in which waste and odd lots are handled which determines whether or not large foci of infestation can develop in neglected produce; and

(4) the size of stacks and the closeness of packing. Most insect species are confined, more or less, to the surfaces of a stack, and weight loss is usually highest at the periphery.

Sometimes, especially in the tropics and in imported produce, materials are infested before stacking. If the core of the stack is infested from the start of storage, then the heat of metabolism of the insects will raise the temperature in the core and, hence, the rate of increase of the insect population and the amount of damage done.

If the stack is small and much of the heat is dissipated, the temperature will remain favourable for insects and weight loss will be very great. If the stack is large, the accumulated heat in the core will get too hot for the insects. After that, weight loss will occur only at the outside of the stack. For this reason, large stacks are advocated in the tropics. However, the high temperature has harmful effects that must be set against prevention of weight loss. Continuous high temperature accelerates chemical degradation, especially for vitamin destruction and the development of rancidity.

Stability of Micro-Ingredients in Fish Feed

Diets for various fish species have undergone rapid development in the last decade (most commercial fish feed is in pelleted forms). It is, therefore, only natural that the effects of mixing and pelleting on the more unstable constituents of dry-type feeds are getting increasing attention.

It is the general practice of fish feed manufacturers to fortify feeds with vitamins, minerals, preservatives, and sometimes pigmenters and antibiotics. These additives are used in such small quantities that they are generally termed micro-ingredients. The use of micro-ingredients in manufactured feed represents an increase in the feed manufacturers responsibilities regarding premixing, product stability, quality control, etc.

Vitamin B_{12}, for example, might be added at as low a level as 5 mg representing a dilution of 5 ppb. If such micro-ingredients were added directly to feed, uniform distribution would be impossible in commercial feed mill practice. The problem becomes even greater when ten or even twenty of these ingredients are added to an individual feed. On the other hand, it also portrays the increased opportunities to the feed manufacturer to render additional service to the fish raiser. Today's typical fish food formulae may contain from 20-30 different micro-ingredients.

Selection of Micro-ingredient Sources

Particle Size: In selecting a commercial source of a micro-ingredient, one must consider carefully the question of particle size. At the usual low levels of supplementation, there are very definite upper limits on particle size if one is to ensure adequate distribution in the finished feed. Minute intakes of very small fish also demand uniform dispersion of all ingredients. However, a correlation exists between stability and exposed surface per unit weight. A balance exists between distribution requirements and ingredient sensitivity which contra-indicates pulverising. When comparing sources of micro-ingredients one should ensure that the average particle size of the active micro-ingredient is optimum in relation to the level of incorporation in the feed and its stability.

Flow Characteristics

Active ingredients should be free-flowing. In some cases a fine particle size is essential and extra-blending is mandatory, followed by light milling to remove balls and lumps. Particle shape should correspond to that of the diluents. Micro-ingredients which acquire static electric charges require extra care in handling. This can often be minimised or eliminated by using different forms of the ingredients where similar choice can be exercised with normally hygroscopic micro-ingredients to avoid caking. Often such tendencies toward moisture pick-up will be satisfactorily reduced in a finished premix, particularly when a suitable conditioning agent is added.

Stability of Micro-ingredients

New problems have arisen with the switch to pelleted feeds for fish. The physical process of pelleting a composite of ingredients has raised many questions concerning the stability of the more labile ingredients and/or nutrients. This concern is increasing as techniques employing more steam, thicker dies, and higher pellet temperatures are attempted.

The term "stability" has been widely used and is sometimes abused. Often it is defined in terms of analogies. One of the most useful states: "Stability is a state of condition of matter which resists change". In applying this definition to the subject of interest, it appears that interest lies in any force or stress, or other action which could conceivably change the nutrient level and/or biological activity in finished feeds.

Before going into the myriad of factors affecting micro-ingredient stability, it will be necessary to distinguish between two types of instability.

It is often concluded that an active ingredient has low chemical stability when, in fact, faulty manufacturing and assay procedures result in apparently low activities in the sample. The number of factors or combination of factors which might cause such mechanical or physical instability are almost limitless. Some of these factors are discussed below.

Considerations of Stability

Separation: Any finished feed which can separate into two or more of its component parts through the action of such stresses as dusting, vibration, screening, and generation of static electrical charges is said to be physically unstable. If this property is not recognised, analytical procedures will indicate irregular and, almost invariably, low assays for one or more active ingredients.

Under-mixing or Over-mixing

If too much time is omitted from the mixing cycle, in homogeneity in the final mix will produce irregular assay results. Over-mixing, on the other hand, leads to actual destruction of sensitive micro-ingredients.

Improper Analytical Procedures

Occasionally, a physically stable and well-mixed feed or concentrate will be submitted to a laboratory for one or more active-ingredient assays. Abnormal values are sometimes reported because old or improper reagents were used, the approved procedure was not followed, or mistakes were made in calculations. This type of error is difficult to detect except through repetition of the assay, change in analytical personnel, and/or change of laboratory conducting the assays.

In each of these aforementioned cases, the first reaction would be to condemn the active ingredient as being unstable. Therefore, it is absolutely essential that all factors of physical or mechanical instability be carefully eliminated before charging such deficiency to chemical instability or biological inefficiency, or both.

Pelleting of Feeds and Micro-Ingredient Stability

The very nature of feed pelleting seems to render it almost immune to precision and standardisation. The operation itself might be considered as variable since it is generally very difficult to duplicate conditions and results from day to day, and sometimes even from run to run.

Major variables of feed pelleting include:

Feed Uniformity

Non-uniformity of pellets and crumbles may be very critical for trout having low feed intakes. Appropriate milling of bulk ingredients and adequate premixing of micro-ingredients are generally helpful in eliminating particle separation. The use of ingredients, such as fish solubles or oils, helps retard particle separation.

Mash Conditioning

Moisture (steam or water) is added in varying amounts to most feeds before pelleting. It is apparent that increasing the moisture content of feed products by steam or water, will accelerate the decomposition of sensitive micro-ingredients. Since this, apparently, is an irreversible process, the use of hygroscopic forms of additives should be avoided as much as possible.

Pelleting Temperatures

Temperatures inside the die significantly exceed those in the hot pellets as they leave the die. Die temperature often exceeds 120°C and extruded pellets may reach 92°C. Elevated temperatures accelerate decomposition of micro-ingredients by increasing,- chemical reactivity.

Animal fats, wax coatings, and several vitamins melt under these elevated temperatures. Many substances are known to be most unstable near their melting point (e.g. vitamin C).

Some factors that influence pelleting temperatures include:

(a) mash composition, fat and mineral levels;

(b) die size, thickness, age, polish;

(c) temperature and quality of steam added;

(d) production rate;

(e) time feed is in the die; and

(f) hardness of the pellets produced.

Pellet Pressures

Pelleting pressures are thought to run between 5 000 and 10 000 psi (300-600 kg/cm^2). The use of such pressure can cause deformation and crushing of certain crystalline micro-ingredients. Wax coatings and stabilised fat beadlets can be obliterated. Reduced particle size increases the exposed surface area of ingredients with resultant greater susceptibility to decomposition. Furthermore, the compression simultaneously places sensitive additives in more intimate contact with other incompatible ingredients.

Feed Time in Die

The next pertinent question is how long the feed is actually in the die and subjected to these severe conditions. Data indicate that with good production rates, the holding time in the die is a fraction of a second.

Diluents for Premixes

After deciding on a source of each of the needed micro-ingredients, attention must be directed to diluents. Any diluent used should maintain or improve the stability and physical properties of the active components (micro-ingredients). Most feed formulators attempt to match the micro-ingredient densities to those of the diluents whenever possible. This is very desirable, although not a completely reliable procedure to eliminate separation.

The shape and surface texture of the micro-ingredient and diluent particles can often be most important. Better premixes can often be prepared by employing a blend of several diluents rather than with just one. It is a rare, but happy, coincidence when only one diluent is required. In most cases, it is necessary to employ a predetermined ratio of three or more diluents for best results. One or more diluents may require pre-milling to obtain a desirable particle size range. The micro-ingredient, depending on its own particle size and density, may be blended before or after diluent milling. Finally, it is generally desirable to employ diluents having definite nutritive value (e.g., calcium carbonate, soybean meal, etc.).

A number of additional factors influence the usefulness of diluents. Most micro-ingredients exhibit optimum stability in the pH range of about 5.5-7.5. The pH value becomes more important with increasing moisture levels in blends. Proposed formulations should be carefully checked for this property. The presence of natural or added fats may create storage problems if rancidity develops due to inadequate stabilisation. The off-flavours and odours resulting may affect palatability, and peroxides and carbonyls can reduce vitamin potencies as well as destroy pigmenters (e.g. oxy- and keto-carotenoids). Trace minerals can be troublesome, especially in the presence of high moisture levels, to many of the water and fat soluble vitamins.

Unprotected fats rapidly oxidise in the presence of trace minerals and moisture to form peroxides which can, in turn, decompose fat-soluble vitamins. It is essential to employ diluents which have low moisture equilibrium and that are relatively non-hygroscopic. If carriers have pronounced hygroscopic tendencies, there may be caking, potency loss, increased degradation due to trace metals, and possible mould growth with attendant overheating during summer storage. Inorganic diluents are generally ideal from the moisture equilibrium standpoint, but they seldom can be used exclusively because of high density. Commonly used organic carriers, such as soybean meal or distillers grains, are often pre-dried. The finished supplements are generally packaged in multiwall paper bags having a vapour and water-proof laminated layer.

A great deal of thought should be devoted to problems involved in handling micro-ingredients. A case in point concerns the potency at which these are purchased. Some of the common B-vitamins can be purchased in almost totally pure forms, for example, niacin. Others are purchased in more dilute forms; for example, vitamin B_{12} Still others are purchased in protected (stabilised) forms to preserve biological activity; for example, vitamin A.

In general, the question of whether to buy ready-to-use premixes, or concentrated ingredients to make one's own premix, is usually resolved on the basis of experience, safety, and economics. If it is decided to purchase the required premixes, these can usually be custom-made so that they can be utilised without further premixing. Generally, the premix should be designed to be used at a rate of five pounds or more per ton of finished feed to avoid mixing problems. In the final analysis, the amount of premix that should be used depends upon the capabilities of the equipment available and the operating procedures employed in a specific feed mixing operation. It is foolhardy to assume that an operation which is successful in one mill will be equally successful in another.

Table 7: Vitamin Loss Due to Processing. Storage, and Water Leaching [1/]

	Ascorbic acid, %	*Pantothenic acid, %*	*Folic acid, %*	*Thiamine, %*	*Pyridoxine, %*
Processing loss [2/]	17	0	0	0	0
Processing and storage loss (6 months)	97	8	5	11	7
Processing, storage and water leaching (10 sees) loss	67	20	17	17	3
Total (%. of original level)	99	26	21	27	10

[1/] Practical-type diet using conventional feedstuffs

[2/] Immediately after processing

Fish Feed Formulation

Feed formulation is essentially applied nutrition. A number of terms and expressions are introduced that will be put to practical use as information is presented on the nature and qualities of various feedstuffs and the information presented on the nutrient requirements of fish. Precise understanding of these terms is essential to their correct application. One must recognise that some of these terms have a built-in error that cannot be escaped. This does not eliminate their usefulness in feed formulation. However, one must appreciate the fact that some are useful approximations of the values and not true values.

The terms that one needs to understand to formulate practical fish diets are: crude protein level; energy level, either expressed as metabolisable energy (ME) or as digestible energy (DE); specific amino acid levels; crude fibre level; and ash level. Since most complete practical fish diets are supplemented with a vitamin premix at levels in excess of the dietary requirement, this category of nutrients will be ignored temporarily.

The potential problems occur when one fails to recognise that all of the above mentioned terms, except ME and DE, represent the quantity or level of a nutrient in the feed as determined by chemical tests on a specific sample of a feedstuff. These chemical tests generally correlate well enough with biological methods of feed evaluation (growth studies, tissue, levels) to be very useful to feed formulators, but they are still chemical tests that are subject to experimental error during nutrient level determination.

For example, the proximate composition of fish meals changes during the spawning season. Generally, the lipid levels increase before spawning and decrease after spawning. This will alter the percent of protein, ash, and carbohydrates in fish meal as the seasons change. Similarly, many plant feedstuffs vary in proximate composition with their stage of maturity at harvest, location grown, and other environmental conditions, such as the weather. Tabled values represent an average value that is usually close enough to the actual value to allow accurate feed formulation. However, one must be aware that assumptions are being made in order to recognise the potential sources of error that may exist.

Metabolisable, energy and digestible energy values are obtained biologically and, thus, should accurately represent the true energy value of feedstuffs to fish. However, ME values may be obtained in different ways (faeces collection methods) and thus may be subject to experimental error. It has recently been reported that the digestibility of feed by rainbow trout was lower at 7°C than at 11°C or 15°C. At 11°C and 15°C body size (18.6 g, 207.1 g or 585.7 g) did not affect feed digestibility. The digestibility of carbohydrate and energy was slightly reduced by meal size in rainbow trout fed at 1.6 percent body weight. Protein and lipid digestibility was not reduced by meal size. Obvious differences exist between fish species in nutrient digestibility, especially in the carbohydrate fraction of feed. Herbivorous and, to a lesser extent, omnivorous fish have longer digestive tracts than do carnivorous fish and are able to obtain more digestible energy from carbohydrates. An awareness of these facts will prevent misuse of ME and DE values.

Each feedstuff in any diet formulation should be present for a specific reason; i.e., it is a good energy source, it is rich in a limiting amino acid, etc. In addition, each feedstuff in a particular diet formulation should be the least costly ingredient available for its particular function in the diet. This leads to another assumption in feed formulation; that is, any nutrient in a particular feedstuff, such as an amino acid, is just as valuable as the same nutrient in any other feedstuff. This allows feed formulators to interchange one feedstuff with another as cost and availability change. Thus, it is assumed that there is no "ideal formulation", but rather an almost infinite number of possible feed formulations that met the nutritional needs of the fish equally well. While this assumption may not be entirely valid and some nutritional judgement must be employed in any feed formulation, it does seem to be valid in most cases. As with the previously mentioned assumption, an awareness of the potential pitfalls involved is necessary for the fish feed formulation so that allowances can be made in diet formulation and problems can be anticipated and avoided.

Balancing Crude Protein Level

In most animal diets, protein is the most expensive portion and is usually the first nutrient that is computed in diet formulation. The energy level of the diet is then adjusted to the desired level by addition of high energy supplements) which are less expensive than protein supplements. The square method is an easy way to determine the proper dietary proportions of high and low protein feedstuffs to add to a feed to meet the dietary requirement of the animal to be fed.

For example, suppose rice bran and soybean meal were available as feedstuffs to prepare a diet for carp that was 25 percent crude protein. A square is constructed and the two feedstuffs are put on the two left corners along with the protein content of each. The desired protein level of the feed is placed in the middle of the square. Next, the protein level of the feed is subtracted from that of the feedstuffs, placing the answer in the opposite corner from the feedstuff. Ignore positive or negative signs.

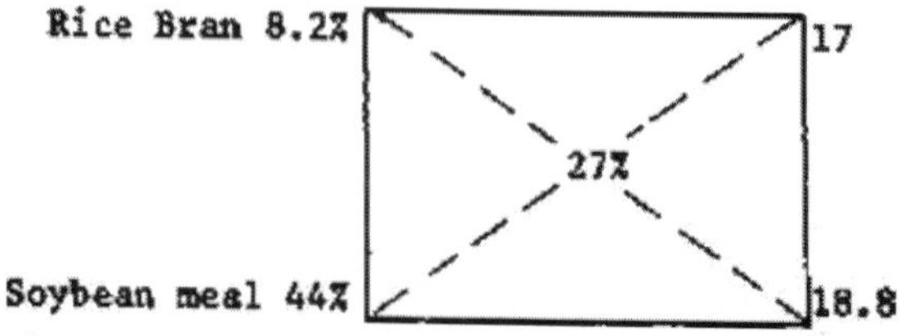

To make the 27 percent crude protein carp feed, we must mix 17/35.8 of rice bran with 18.8/35.8 soybean meal.

Rice Bran 17/35.8 = 47.5%

Soybean meal 18.8/35.8 = 52.5%

So to make 100 kg of this feed we must mix 47.5 kg of rich bran with 52.5 kg of soybean meal.

If more than two feedstuffs are used in a feed, they may be grouped into basal feeds (CP < 20 percent) and protein supplements (CP > 20 percent), averaged within each group, and plugged into the square method. For example, suppose shrimp meal and corn were also available for the carp feed mentioned above. The crude protein levels of the shrimp meal (52.7 percent) and of corn (10.2 percent) are averaged with soybean meal and rice bran, respectively.

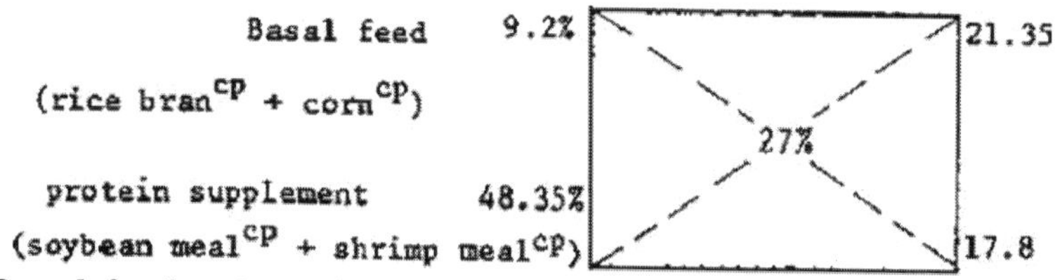

Basal feed = 21.35/39.15 = 54.53%

Protein supplement = 17.8/39.15 = 45.47%

Thus, to make 100 kg of this feed one would mix the following:

Rice bran	27.265 kg
Corn	27.265 kg
Soybean meal	22.735 kg
Shrimp meal	22.735 kg

The square method is helpful to novice feed formulators because it can get them started in diet formulation without the need to resort to trial and error. The square method can also be used to calculate the proportion of feeds tuffs to mix together to achieve a desired dietary energy level as well as a crude protein level. If one wanted to make a feed containing 2 500 kcal ME/kg using wheat middlings

(1663 kcal ME/kg) and anchovy fish meal (4 371 kcal, ME/kg) a square could be constructed as follows:

Steps in Feed Formulation

The first step in diet formulation is balancing the crude protein and energy levels. This can be accomplished by trial and error, by the square method for either crude protein level or energy level and then adjusting, or by solving simultaneous equations. At first, it is helpful to use at least three feeds tuffs during the initial balancing of protein and energy levels: one high in protein and high in ME, one low or intermediate in protein and high in ME, and one low or intermediate in both protein and ME. Once practice makes one more proficient at diet formulation any number of feedstuffs can be used. One must remember to reserve room in the formulation for any feed additive, such as a vitamin or mineral pre-mix.

The second step in diet formulation is to check the levels of indispensable amino acids in the formulation to be sure the dietary levels meet the requirements of the animal to be fed. The requirements of fish for indispensable amino acids is expressed as the dietary level (as a percent of the diet) or as a percent of the dietary protein level. To convert an amino acid level from the percent of diet to percent of protein, divide the dietary level of each amino acid by the dietary protein level. It might be of interest to calculate the dietary levels of all of the indispensable amino acids, but it is not practical to do it all of the time. If the levels of arginine, lysine, methionine, and tryptophan meet the dietary requirements of the fish to be fed, the levels of the other six indispensable amino acids will most likely be above required levels. When using unconventional protein supplements, the levels of all ten indispensable amino acids should be checked.

If the diet formulation is low in any amino acid, a feedstuff that contains high levels of that amino acid must be added to the diet at the expense of another ingredient. Once the amino acid requirements are met, the dietary protein and energy levels must be rechecked to, see if any substitution of ingredients has imbalanced the formulation.

A diet mixing sheet should be constructed to standardise diet formulation. The amino acids listed are for illustration purposes only and may be changed to suit different circumstances.

In practical feed formulation, pellet quality and acceptability must be considered in addition to nutrient levels and cost. These considerations will vary from species to species and with the type of pellet being made, and are dealt with in other sections of this manual.

Best-buy Techniques

The price of the feedstuffs used in diet formulations must be considered to formulate a cost-efficient diet. Feedstuffs can be compared with one another on the basis cost per unit of protein, energy, or amino acid. For example, suppose one has wheat middlings and wheat millrun available for a fish diet, which feedstuff would be the least expensive source of energy?

Wheat millrun costs US \$ 0.0858/kg, and contains about 1200 kcal ME/kg.

$$\text{Cost/kcal}=\frac{.08758}{1200}=\text{US\$0.0000715/kcal}$$

Wheat middlings cost US \$ 01883/kg and contain 1663 kcal ME/kg.

$$\text{Cost/kcal}=\frac{0.1883}{1663}=\text{US\$0.00001132/kcal}$$

Thus, the wheat millrun which has a lower ME value for fish is the better buy because it costs less per kcal.

To compare oat groats and wheat middlings on a cost per unit ME basis one would do the following:

Wheat middlings = US\$ 0.0001132/kcal, and

Oat groats cost US \$ 0.2652/kg, and contain about 2450 kcal ME/kg.

$$\text{Cost/kcal}=\frac{0.2652}{2450}=\text{US\$0.001082/kcal}.$$

Oats groats, although costing more than wheat middlings, constitute a better buy on an energy basis.

The cost of protein is often the greatest part of the cost of a fish diet. Therefore, substantial savings can be made by using best-buy techniques to determinate least expensive protein supplement. To compare anchovy meal and herring meal, the following calculations are made:

Anchovy meal costs US \$ 0.5357/kg, and contains 70.9 percent protein.

$$\text{Cost per kg protein}=\frac{0.5357}{0.709}=\text{US\$0.7556/kg protein.}$$

Herring meal costs US $ 0.4709/kg, and contains 76.7 percent protein.

$$\text{Cost per kg protein} = \frac{0.4709}{0.767} = \text{US\$}0.61395/\text{kg protein}$$

On the basis of cost per unit protein, herring meal is less expensive as a dietary ingredient than is anchovy meal.

To compare feedstuffs on the basis of cost per unit of an amino acid, one can calculate the best buy in the same way as before.

For example, sesame oil cake which has twice as much methionine content as does groundnut cake on a per unit protein basis would be a more attractive buy at comparable prices.

These kinds of comparisons are only valid if the nutrient in one feedstuff is as valuable or available to the animal as the same nutrient in another feed. Such comparisons should be made whenever prices change.

Chapter 3

Material Flow in Feed Manufacturing

The technology of feed processing has undergone substantial improvement in recent years. It was only sixty years ago that feeds tuffs were mixed on the warehouse floor by the use of a shovel. Feed processing has progressed from the simple mixing of several ingredients by hand to mechanical mixing, to continuous mixing, and now to computer controlled mixing and pelleting. However, the basic concept of mixing ingredients together to result in a nutritionally balanced feed, has remained unchanged.

To accomplish the mixing of different ingredients, grinding these ingredients to similar particle sizes, and then putting them together in a single unit, requires a considerable amount of specialised equipment and technical expertise. Some feed plants are designed for specific functions, such as making poultry feeds exclusively; others are designed for producing a variety of feeds. Regardless of the specific purpose of a feed plant, material flow follows a basic pattern.

Receiving

The first operation in the feed processing plant involves the receiving of raw materials into the plant premises. Feed ingredients arrive in sacks, or other small containers, and in bulk.

Sacked ingredients are checked for identification and condition. They are then logged in after segregation of drugs and medications. Sacked ingredients must then be stored in a dry location with proper protection from rodent and insect infestation. Sacked stocks are then rotated to minimise staleness, product degradation, and insect infestation.

Bulk ingredients are handled according to their physical form. Liquid ingredients, such as oils and molasses, are generally stored in bulk tanks. Proper storage temperature is maintained and the filter screens are checked periodically. Solid bulk ingredients such as grains, oil meals, etc., are cleaned with a scalper to remove foreign material prior to storage in bins. Bin temperatures are monitored to prevent heating due to grinning respiration.

Processing

Material flow during processing includes:

(a) particle size reduction,

(b) premixing,

(c) mixing,

(d) pelleting, and

(e) sacking.

Coarse ingredients pass over a permanent magnet which removes tramp metal and then through a hammer mill which reduces particle size to the desired screen analysis. Ground material is monitored periodically to ensure size uniformity and to help detect wear of hammermill screen and hammers. The ground material is then routed to ingredient holding bins.

There are two mixing operations in feed milling. One is for the mixing of micro-nutrients; the operation is generally termed pre-mixing. The other mixing operation involves the actual blending of all components of the diet. Micro-nutrients, such as vitamins and trace minerals, are accurately weighed with carrier material which has a density approximating that of the predominant micro-ingredient. The materials are then mixed in a batch mixer for a period of time specified by the equipment manufacturer to ensure homogeneity. The premix is finally routed to the premix holding bin.

Diet mixing begins when augers are set in motion to deliver; the correct amounts of each ingredient including the premix, according to the formula, into the mixer. Where manual changing of the mixer is done, ingredients are weighed out in sacks or hopper carts. The mixing period is according to the equipment manufacturer's specifications, but final mix is checked periodically with a tracer to ensure homogeneity of the mix. If the mixed diet is to undergo pelleting, it is routed to the pelleting bin.

Mixed feed mash for pelleting is first conditioned with steam in the steam conditioner section of the pellet mill, after which it enters the die where it is finally extruded. Freshly extruded pellets are hot and contain excess moisture which is removed during passage through the cooler. Fines are then screened from the cooled pelleted feed and returned for repeating. Fish oil, if added, is now applied prior to the routing of the finished pellets into the packer bins.

Packaging

Most fish feed pellets are sacked. The sacking operation includes: weighing, sacking, taping, coding, and sewing. The sacked bags are then sent to the warehouse for distribution.

Storage and Distribution

Bulk products are stored in large bins. Feed is loaded into specially constructed bulk carriers and delivered to the customer. Feed is discharged from the truck into customer storage facility by means of an auger system, paddle conveyor, or pneumatic system.

Feed Milling Processes

Feed manufacturing and the associated quality control programme are keys to successful fish culture. Unless the fisheries biologist understands and specifies the activities of the feed mill and its laboratory, profitable fish farming will be a matter of chance.

Dry feeds may be ground, sifted, screened, mixed, compressed, expanded, texturised, coloured and flavoured. By one or more of these processes, a wide variety of ingredients can be prepared into a standardised product. Since most fish have size and texture preferences and often react to colour, odour, and flavour, processing research is an integral part of fish culture.

Grinding

Grinding or particle-size reduction is a major function of feed manufacturing. Many feed mills pass all incoming ingredients through a grinder for several reasons:

(a) clumps and large fragments are reduced in size,

(b) some moisture is removed due to aeration, and

(c) additives such as antioxidants may be blended.

All of these improve the ease of handling ingredients and their storability. There are other reasons for grinding and the associated sieving of ingredients in formula feeds before further processing.

Small fish and fry require plankton-size feeds available in dry form as a meal or granule. Extremes in particle sizes are wasteful and often dangerous. Fry have been killed because of their inability to pass through the digestive system large pieces of connective tissue and bone present in dry animal byproducts, or hull fragments found in cottonseed meal and rice bran. On the other hand, dust or "fines" may become colloidal suspensions in water, so dilute that several mouthfuls carry little nutritive value.

The grinding of ingredients generally improves feed digestibility, acceptability, mixing properties, pelletability, and increases the bulk density of some ingredients. It is accomplished by many types of manual and mechanical operations involving impact, attrition, and cutting.

Hammermills

Hammermills are mostly impact grinders with swinging or stationary steel bars forcing ingredients against a circular screen or solid serrated section designated as a striking plate. Material is held in the grinding chamber until it is reduced to the size of the openings in the screen. The number of hammers on a rotating shaft, their size, arrangement, sharpness, the speed of rotation, wear patterns, and clearance at the tip relative to the screen or striking plate are important variables in grinding capacity and the appearance of the product. Heat imparted to the material, due to the work of grinding, is related to the time it is held within the chamber and the air flow characteristics. Impact grinding is most efficient with dry, low-fat ingredients, although many other materials may be reduced in size by proper screen selection and regulated intake.

Most hammermills have a horizontal drive shaft which suspends vertical hammers but for some ingredients, such as dried animal byproducts, a "vertical" hammermill is more efficient. In this mill, the drive shaft is positioned vertically and screens and hammers are positioned horizontally. Material successfully reduced in size to the diametre of screen holes or smaller, are carried by gravity outside the mill and thence by air or conveyor to storage in "make-up" bins. Over-size particles, not easily broken, drop through the mill and may be re-cycled or discarded. Thus foreign materials, such as metal and stones, are discharged before they are forced through the screen causing damage.

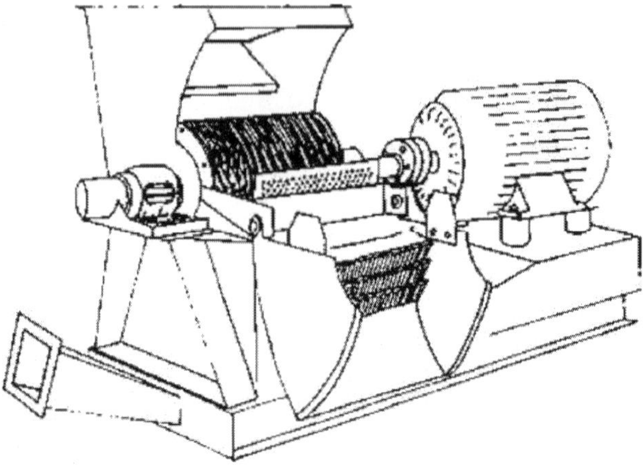

Figure 1: Hammer Mill

Attrition Mills

Attrition mills use the hammermill principle to a certain extent; i.e., shattering by/impact. However, they also impart a shearing and cutting action. Grinding is done between two discs equipped with replaceable wearing surfaces. One or both of these discs is rotated; if both, they rotate in opposite directions.

When one disc is rotated, and the other stationary, the assembly is used for shredding and deferring. Often materials which have been coarsely ground by other mills, are passed through an attrition mill for blending or smoothing out an ingredient or mixture containing liquids which may have clumps. The discs of an attrition mill are generally in a vertical position so that materials not capable of reduction can pass by gravity out of the grinding area.

Roller Mills

A combination of cutting, attrition, and crushing occurs in roller mills. These are smooth or corrugated rolls rotating at the same speed set at a pre-determined distance apart with material passing between the two. A tearing action may be added by operating the rolls at different speeds and by corrugations which are different for each roll; i.e., the top roll may have off-radial spiral corrugations and the bottom roll lateral corrugations.

This last type, called a "Le Page cut" is used in making granules from hard pellets, as it provides a breaking surface without much impact to cause dust. Roll grinding is economical but limited to materials which are fairly dry and low in fat.

Cutters

Rotary cutters are a type of grinder which reduces dry particle solids mainly by shearing with knife edges against a striking plate. The mill also includes the processes of attrition and impact, although these actions are limited if the material is easily reduced by cutting and the screen limiting discharge has large perforations. The mill consists of a rotating shaft with four attached parallel knives and a screen occupying one fourth of the 360 degree rotation. The mill is best used to crack whole grains with a minimum of "fines". It is not used as a final process for reducing the size of ingredients used in fish feeds.

Screening

Associated with grinding feeds for fish fry, a sieving system is required which classifies materials to any desired particle size. The "overs" in this system may be re-ground or rejected. The "throughs" may be selected to comply with fish preferences for size and mixed according to formula specifications. Feeds sifted through a 177 micron opening (a U.S. No. 100 sieve) have been successfully used for increasing survival and growth of minnows and catfish fry. Hammermill or impact grinding of dry feeds, especially cereal grains, creates particles within the range called "dust", and a dust-collecting system may be necessary to remove this. An excess of dust in the feed may lead to gill disease, a situation where organic matter adhering to gills becomes a nutrient for bacteria or parasites.

The problem of excess dust formed by grinding feeds may be partly alleviated by adding a spray of oil or a semi-moist ingredient, such as condensed fish solubles or fermentation solubles, on feeds entering the grinder. Dehydrated alfalfa is prepared as a dust-free meal, similar in texture to a sifted crumblised pellet, by spraying mineral oil into a hammermill chamber during grinding.

Mixing

The objective of feed mixing is to start with a certain assortment of ingredients called a "formula", totalling some definite weight. This is processed so that each small unit of the whole, either a mouthful or a day's feeding, is the same proportion as the original formula. Mixing is recognised as an empirical unit operation, which means that it is more of an art than a science and must be learned by experience.

Feed mixing may include all possible combinations of solids and liquids. Within each ingredient are differences in physical properties. For solids there are differences in particle size, shape, density, electostatic charge, coefficient of friction as represented by the angle of repose, elasticity or resilience and, of course, colour, odour, and taste. For liquids there are differences in viscosity and density.

The term "mixed" can mean either blended, implying uniformity, or made up of dissimilar parts, implying scattering. As applied to formula feeds, the objective of mixing combines each of these definitions; i.e., the scattering of dissimilar parts into a blend. However, it is improbable that uniformity is attained with particles within a, sample arranged in some order of position or concentration. That is only a quality control; goal. It has been suggested that a proper title for a discussion of mixing should be "mixing and unmixing", for during the operation there is a constant tendency of particles which have been mixed to become separated. Three mechanisms are involved in the mixing process:

(a) the transfer of groups of adjacent particles from one location in the mass to another,

(b) diffusion, distribution of particles over a freshly developed surface,

(c) shear, slipping of particles between others in the mass.

In theory, the position of particles within a container is determined by chance, and the effects of chance accumulate until they outweigh the direct effects of mixing action. In the mixing of liquids, chance movement of components creates order or uniformity. With dry solids, chance distribution creates disorder. When disorder is at a more or less stable maximum, it may be called "random". Many factors in dry solids cause particles to avoid a chance or random arrangement. In fact, the result of mixing feed ingredients may be a definite pattern of particle segregation or non-random arrangement.

Particle segregation is due to differences in the physical properties of ingredients and the shape and surface characteristics of the mixer. Particle size may be the most important factor in causing segregation. An improvement in mixing which approaches random distribution of solids by decreasing particle size can be measured quantitatively by statistical methods. In general, the smaller and the more uniformally sized the ingredients are prepared, the more nearly they will approach random distribution during mixing.

In many formulae, a decrease in particle size is necessary to attain a sufficient number of particles of an essential additive (vitamin, mineral, medication) for dispersion in each daily feed unit.

This may require the particle size to be the diametre of dust, 10 to 50 microns. Certain ingredients are unstable in finely divided form and likely to acquire an electrostatic charge.

Concentration of particles on a charged surface, roughness of the mixed and stickiness of oily and wet ingredients are factors in causing segregation when very small particles are mixed and when these are much smaller than the bulk of other ingredients.

Mixing may be either a batch or a continuous process. Batch mixing can be done on an open flat surface with shovels or in containers shaped as cylinders, half-cylinders, cones or twin-cones with fixed baffles or moving augers, spirals, or paddles.

Continuous mixing proportions by weight or volume, is a technique best suited for formula feeds with few ingredients and minimal changes.

Horizontal Mixers

Continuous Ribbon Mixers: The continuous or "twin-spiral" mixer consists of a horizontal, stationary, half-cylinder with revolving helical ribbons placed on a central shaft so as to move materials from one end to the other as the shaft and ribbon rotate inside. Capacity can be from a few litres to several cubic metres.

The speed of shaft rotation will vary inversely as the circumference of the outer ribbon; usually optimum between 75-100 metres per minute. Since material travel is from one end to the other, either end may be used for discharge. These mixers may be inverted for cleaning.

Non-continuous Ribbon Mixers

Non-continuous or interrupted ribbons are similar to the continuous ribbon mixers except that short sections called "paddles" or "ploughs" are spaced in a spiral round the mixer shaft.

Action is different from that of continuous ribbon mixers, and may be more satisfactory for mixing liquids with dry solids.

These mixers are made in a wide variety of sizes with travel of the outer diametre of paddles from 100 to 120 metres per minute.

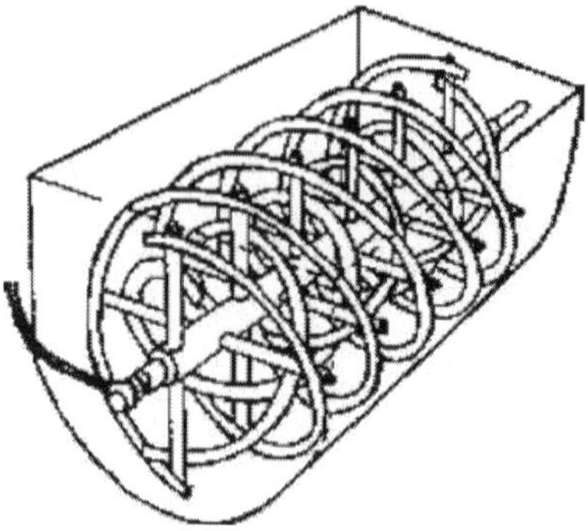

Figure 2: Hammer Mill

Vertical Mixers

Vertical mixers may consist of a cylinder, cone, or hopper-shaped container, with a single or double screw (auger) located vertically through the centre. The screw operates at speeds of 100 to 200 rpm and vertically conveys incoming materials from the bottom (generally the intake) end, like a screw conveyor, to the top where they are scattered and fall by gravity. This sequence is repeated several times until a blend is attained (usually from 10 to 12 minutes). These mixers may also be loaded from the top. Results show that vertical mixers are not efficient for uniform mixing of solids and liquids or for materials of quite different particle size or density. This unit is difficult to clean and there may be inter-batch contamination.

Other Types of Mixers

A third type of mixer is the horizontal revolving drum. This can be a straight-sided cylinder or a cylinder tapered at each end. The sides may be smooth or fixed with baffles or shelves to pick up and drop ingredients. Smooth, dry materials of uniform physical properties are blended best in this type of mixer.

A modification of this type is the turbine mixer which is a fixed cylinder with revolving shaft to which are fixed paddles, ploughs, scrapers, or shelves designed to re-pile materials continually. This mixer is often used as a cooker to dry fish wastes and to blend various types of fish meal into a standardised product. They are also particularly efficient for mixing heavy ingredients and for adding liquids to mixtures which would clump or cake in another type of mixer. Some particle size reduction (grinding) may occur on soft materials, such as rice bran and alfalfa leaves. A complete mixing can usually be attained in 3 to 6 minutes unless longer time is necessary to eliminate lumps caused by added liquids. Mixer shaft rotation is regulated to provide some centrifugal action, but this must not be excessive.

The "Nauta" mixer originated in Holland and is constructed in the form of an inverted cone with a mixing screw inside rotating around the inside wall. The mixer is made in a variety of sizes from laboratory models, for premixing chemical and vitamin additives, to very large production sizes. It is excellent for premixing trace elements and works very well for adding moderate amounts of liquids into dry ingredients.

Another type of mixer called the "entoleter" consists of a high-speed rotating disc which throws the ingredient charge with considerable force against the walls of a chamber. This mixer functions well to smooth out clumps or balls of compacted ingredients and will cause eggs of grain weevils to become inactive. However, since it may shatter vitamin A beadlets encapsulated in gelatine, it is not recommended for all mixtures.

Liquid Mixers

Oils and water-miscible oil preparations are often added to dry ingredients as a source of energy or as a specific nutrient. Although the oil-soluble vitamins. A, D, E, and K, are available in dry carrier concentrates, they may be obtained in pure form and premixed by the feed manufacturer. Liquids containing nutrients can be mixed faster and with more uniformity than the same nutrient in dry concentrate condition. Therefore, a liquid blender may be needed in the feed plant.

Liquid blenders usually consist of a horizontal tub or cylinder with a number of wires or paddles equally spaced around a shaft which revolves inside. Sometimes the shaft is hollow and liquids are forced through holes in the paddles in a spray effect. Some models have a shaft speed of 400 to 600 rpm while others rotate at 1 200 rpm. Ingredients such as condensed fish or fermentation solubles, molasses, or fish oils are often premixed in a bowl type variable speed mixer, blending the liquid with dry ingredients.

Mixing Operation and Evaluation

Accurate mixing requires the addition of ingredients in a tested sequence from batch to batch. The usual practice is to add large-volume ingredients first, then those of smaller amount. Unless already premixed, liquids should be added after all dry ingredients have been mixed. Total mixing time is critical and is influenced by the composition of the formula.

All mixers should be calibrated by laboratory recovery of known additives (physically or chemically) so that under and over mixing does not occur. Uniformly sized salt, graphite, or iron particles coated with water soluble dyes are often used as "tracers". Each mixer should be calibrated for its mixing time and capacity by volume for best results.

Pelleting

The transformation of a soft, often dusty feed into a hard pellet is accomplished by compression, extrusion, and adhesion. The general process involves passing a feed mixture through a conditioning chamber where 4 to 6 percent water (usually as steam) may be added. Moisture provides lubrication for compression and extrusion and in the presence of heat causes some gelatinisation of raw starch present on the surface of vegetative ingredients, resulting in adhesion.

Within 20 seconds of entering the pellet mill, feed goes from an air-dry (about 10-12 percent moisture) condition at ambient temperature, to 15-16 percent moisture at 80-90°C. During subsequent compression and extrusion through holes in a ring' die, friction further increases feed temperature to nearly 92°C. Pellets discharged onto a screen belt of a horizontal tunnel drier or into a vertical screened hopper are air-cooled within 10 minutes to slightly above ambient temperatures and dried to below 13 percent moisture.

Contrary to early belief, finished pellets contain practically all the nutrients found in feedstuffs and additives as compounded. The loss of thermolabile vitamins used in additives, which may be slight or extensive in the case of vitamin C, may be compensated for by extra supplementation of these in the vitamin premix to comply with formula requirements. Diastatic enzymes (alpha and beta amylase) present in whole grains and cereal byproducts are still active after processing by grinding and pelleting, although powdered enzymes added as an ingredient are inactivated.

Application

Mechanically, the process of pelleting involves forcing soft feed through holes in a metal ring-type die. These holes may be round or square, tapered or non-tapered. Single or double rolls mounted inside the die ring on a cam or eccentric, turn on a rotating shaft as friction develops (due to the presence of feed between roll and die). Feed is forced through the die holes in increments so that dissection of a finished pellet shows tight layers of feed mixture. The die is driven by a motor and the rolls turn only as feed between rolls and die develops friction.

To make dry feed particles pliable for close compression and to decrease friction and absorb mechanical heat, water is often added to the feed, either as the formula is mixed or in the conditioning chamber of the pellet mill. Water may be provided as liquid and/or vapour. If water is provided in the form of steam, two objectives are accomplished:

(a) a high volume of water vapour condenses on the surface of feed particles, wetting and softening them for better compression, and

(b) the high temperatures of this steam cause some gelatinisation of raw starch present in all vegetative ingredients, providing the necessary adhesion for firm pellets.

If sufficient moisture cannot be added as steam, pretreatment with water may be used to gain the desired lubricating effect.

It is not absolutely necessary to add steam to a soft feed in order to compress it into a pellet. Materials such as rice bran, ground cottonseeds, and palm kernel cake may be pelleted with no added moisture. The resulting pellets are often slightly charred from high temperature and the electrical energy consumption is much more than would be needed if moisture were added. The high fat content of these materials provides lubrication, but this does not soften the fibre sufficiently to avoid excessive heating caused by friction. The ring die and rolls of a pellet mill exposed to high temperatures show metal fatigue and must be replaced often. The proper conditioning of dehydrated alfalfa meal will permit the manufacture of over 2 000 tons of pellets during the life of a standard die of 40 mm thickness.

Overall, the texture of a soft feed mixture is changed from a meal-like material with bulk density approximately 0.4 g/cc, to a pellet with bulk density of 0.5 - 0.6 g/cc. Within the ring die, pressures of 75-600 kg/cm^2 are attained. Feed mixtures containing large amounts of fibrous ingredients often result in pellets too hard for gastric breakdown and digestion in fish. On the other hand, high-fat feeds and an excess of moisture cause poor quality pellets. Pellet quality may be defined as a certain hardness or water stability which assures efficient use without loss in handling on land or in water.

Feed formulation and operation of the pellet mill may be balanced to supply fish with a feed that is acceptable, available, and easily digested. The inter-dependent variables present in ingredient selection are subjects for research in each area of fish culture and for each fish species. Variables resulting from mechanical operation of the equipment may be listed here:

(a) Pellet die thickness as related to diametre of hole is a factor in pellet quality.

(b) Speed of rotation should also be considered for each die thickness/hole diametre combination.

(c) The speed at which feed is introduced into the feed-conditioning chamber affects the moisture/temperature relationship which in turn relates to pellet quality.

(d) The amount of steam added to a given volume of feed should be in balance such that the drive motor on the pellet mill is operating at its maximum amperage. The opening of the steam valve at the pellet mill has a direct relation to the amount of water entering the feed as steam at any given steam pressure.

(e) Atmospheric conditions in the factory, especially relative humidity, which pre-condition the ingredients before processing, will affect die selection and operational settings.

All of the above items must be examined for each feed formula, which shows that pelleting is more of an art than a science. It should be emphasized again that the pellet mill operates most efficiently when the motor amperage use is optimum for the voltage available. It is important to watch the ammetre gauge frequently during pelleting. At the start of a pelleting operation, a small amount of soft dry feed enters the pellet chamber for compression and the gauge will respond to the load with less than optimum results.

The addition of steam results in a lowered reading, showing the lubricating effect. More dry feed may now enter the conditioning chamber and, as it reaches the die, the ammetre will move to a higher reading. Additional steam will lower this. By adjusting the dry feed intake and the steam valve opening, a stable condition will result where the motor is operating at its maximum rated amperage and pellet production will be at maximum capacity. These operational conditions normally coincide with maximum pellet quality afforded by the composition of the feed mix.

Although power transmission by means of a V-belt drive is shown (1), gear drive coupling of motor to the main shaft (2) is also possible. Feed from a surge bin (3) is metred into the steam conditioning chamber by a variable speed screw feeder (4). Paddles shown in the steam conditioning chamber (5) agitate the feed to ensure even blending of feed and steam. The conditioned feed is then fed by means of a distributor auger (6) into the pelletizing chamber (7) where it is extruded through the die (8). As the die rotates, feed is pressed against its inner wall by a set of 3 rollers (9). Due to its high rotational speed and heavy load, the die has to be securely harnessed to the shaft. This is achieved by means of a two-piece die cover assembly (10) and 12 strong bolts (11). The rollers usually have a hardened shell with indentations on the surface to provide traction and to reduce slippage of feed. Feed within the pelletizing chamber is continuously redistributed by adjustable feed ploughs (12). Extruded pellets of appropriate lengths are cut off by an assembly of knives (not shown) mounted on the inside of the die casing.

A control valve introduces dry steam into a header from which, through several port-holes, steam enters the conditioning chamber in contact with dry feed. Between this valve and the steam generator or boiler are a strainer and trap to remove condensate, providing only dry steam in the mill. At the discharge end of the conditioning chamber is a gate to restrict feed from immediately leaving and allows more time for moisture to be absorbed into the feed. A chute or funnel usually guides moisture-conditioned feed into the pellet chamber where compression and extrusion occur. If the ammetre goes much above the optimum reading, this chute may be quickly raised to prevent a choke-up of feed in the die holes. From the pellet chamber, feed may be directed by means of a butterfly valve to the cooler or on the floor for inspection.

Another variable that may be introduced into the operation of a pellet mill is the rotational speed of the die. For the production of small diametre pellets (i.e., 3 mm or less) high rotation speeds are used. This results in a thinner layer of soft feed inside the die ring ahead of the rolls, and for a given volume of feed the efficiency of pelleting and pellet hardness are improved. Die speeds may be changed by replacing the pulley on the main motor shaft of the pellet mill. Speeds generally range from 130-400 rpm. Feeds-of low bulk density are formed best in dies rotating at higher speeds.

Influence of Feed Composition

Additional factors known to influence pellet quality include bulk density of the soft feed, its texture, chemical composition (fat, fibre, carbohydrate, protein, and moisture) and prevailing ambient conditions of temperature and relative humidity. Grinding improves pellet quality by reducing air spaces between particles, allowing closer surface to surface contact for a given volume of feed; i.e., it increases bulk density. Large pieces of any ingredient in a feed formula result in weak spots in the pellet, especially if these are fibrous or bony. Grinding also increases the total surface area of a given weight of feed, thereby allowing more space for steam condensation during the conditioning process.

This results in a higher feed temperature and more water absorption which together, within the time available, increases gelatinisation of raw starch. Starch gelatinisation has been mentioned previously (Section 3), but since it is the key to water-stable pellets, further discussion may be beneficial.

Starch gelatinisation is the rupture of starch granules, thereby allowing the linear and cyclic molecules to hydrate and become sticky in the presence of water. Gelatinisation occurs by mechanical means such as grinding, pressure, and by hot water. Soft feed at an environmental temperature of 25°C can be brought to a temperature of 85°C by the addition of 4 to 6 percent moisture from steam. Frictional heat due to passage of feed through the pellet mill adds 2 or 3 degrees of temperature. The moisture, temperature, and time involved combine to impart a sticky surface to starch-containing ingredients which, when subsequently dried, improved pellet hardness and water stability.

Since fat and moisture lubricate the soft feed as it passes through holes in the ring die, reducing friction of compression and extrusion, a certain level of these are necessary for pelleting. If the total crude fat in a feed formula is much more than 8-10 percent or the moisture is much above 15 percent, too much lubrication is provided and a poor quality pellet in terms of hardness results. Dies may be designed to accommodate high-fat feeds by tapering or slanting the Holes or by increasing the hole depth in relation to its diametre. Dietary fat in the final feed may be increased beyond that in the formula by spraying liquid fat or oil onto pellets or crumbles after they leave the pellet mill. An external coating of 5-6 percent stabilised fat does not appreciably soften the pellet except externally, and this may be a lure both in odour and texture for certain fishes.

In some countries, high humidity may cause problems in pelleting. Sun-dried ingredients and those which absorb moisture on storage (i.e. marine products contain sodium chloride which is hygroscopic) may need to be balanced with low-fat, fibrous ingredients or limited in their use. Low-fat cereals or cereal by-product cassava cake and pressed copra may be used in high-moisture, high-fat feeds to provide resistence to compression and extrusion (necessary to form a hard pellet). If the formula contains sugar or molasses, low-frictional heat is an asset, for carmelisation occurs at about 60°C.

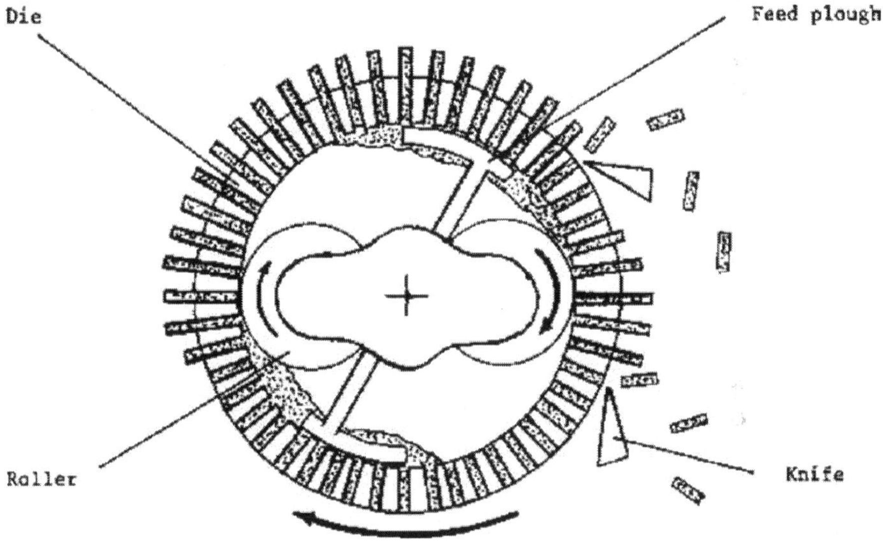

Figure 3: Pellet extrusion (arrows indicate direction of rotation of moving parts)

Cooling and Drying

The temperature imparted to pellets in the process of their manufacture assists the removal of moisture by the air-drying process. Generally, within ten minutes after extrusion, hard pellets are cooled to ambient temperature and brought to a moisture content slightly above that of the entering soft feed. This may be done by spreading pellets in a thin layer on the floor and blowing air over them. Commercially, it is done by passing the hot pellets through a vertical or a horizontal chamber designed to bring air at ambient temperature into intimate contact with the outer surface of the pellets.

Vertical Cooler-dryer

Moisture added as steam, provides a large volume of lubricating material with a small weight. In addition, the temperature imparted to pellets by steam is a major factor in their subsequent drying. If air at ambient temperature is drawn or blown through the newly formed pellets, its ability to take up water depends on an increase in temperature.

Air has an avidity for moisture directly related to its temperature. Air passing over hot pellets will increase in temperature and thus increase its water-holding ability. If cold or warm water were used to lubricate soft feed before pelleting, the final pellet temperature would not reach that necessary to gelatinise raw starch or significantly increase the ability of air to remove moisture in the cooler. Heated air may be used in the cooling-drying process, but this is generally uneconomical. Heating newly-formed, wet pellets on trays in an oven to attain drying does, however, result in very water-stable feed.

Pellets are discharged from the mill into the top of a flat-sided hopper and dropped into an attached cooling bin. This is divided in the middle with a plenum connected to the suction side of a blower fan. The weight of pellets filling the cooler pivots perforated louvers on the two sides to allow cool air to permeate the hot pellets, removing moisture, and cooling the pellets before entering the plenum for discharge through the blower. Pellets leave the bin at the bottom via discharge gates at a rate regulated by the amount of hot pellets entering the cooler. This ensures uniform cooling and drying of pellets.

Pellets gradually fill the cooler section of this system until they reach a diaphragm located near the top of the enclosed hopper. Pressure on this activates the discharge gates and the pellets are discharged until the level is below the diaphragm. A bucket elevator or drag screw conveys cooled pellets to either a storage or bagging bin or to a truck for bulk delivery. This type of cooler is preferred when space is limited and is generally more economical than the horizontal type.

Horizontal Cooler-dryer

Coolers of the horizontal type consist of a moving wire belt or sectional belt of perforated metal trays which convey pellets from the discharge spout of the pellet mill. The depth of pellets on this belt and its speed of travel may be adjusted so that pellets leave for storage at a desired moisture and temperature. Horizontal coolers may be a single deck with pellets discharged at the end opposite the intake, or a double deck with two belts in the same enclosure; pellets return to the same end as they entered. Air from a centrifugal fan is made to flow from the cooler bottom through the layer of pellets. As in the vertical cooler, air is discharged into a dust-collecting system which removes the "fines" or particles which separate from the pellets. These are returned to the mill continuously for re-pelleting.

Crumbles

Cooled pellets may be ground on corrugated rolls and the resulting product sifted into various sizes of granules or crumbles. For small fish, the physical properties of crumbles are often more desired than a meal ration and easier to manufacture than a small pellet. Crumbles provide a multi-faceted surface to reflect light, this being a lure for sight feeders. They ensure that all the ingredients of a formula will be ingested, whereas components of meal feeds separate on entering water, allowing selection of certain ingredients, and the-solution or colloidal suspension of others.

Production of pellets of 4 mm or more in diametre proceeds at a higher rate than when feed is compressed into a smaller diametre. Pellet dies of small diametre holes are expensive and the time consumed in changing dies can make the manufacture of small amounts of fingerling feed quite costly. Generally, crumblising rolls should be complementary equipment in any pelleting operation.

Screening or Grading

Some sifting is necessary in the production of pellets and crumbles. Small fragments (fines) are produced as hot, moist pellets are cut off from the die inside the pellet chamber, and as pellets pass through the cooling and conveying equipment. Fines may be returned to the pellet mill for reprocessing or used as feed for fry.

There are many types of sifting and grading systems both manual and mechanised. Most shake or rotate from side to side with material passing over screens of specified openings and covered to confine dust to the equipment. Sifting is the last process in manufacture of pellets and crumbles and this equipment should be located just above the bagging or final discharge bin.

Use of Hard Pellets

Test feeding in dirt-bottomed ponds or on feeding platforms with minnows, goldfish, catfish, trout, carp, and buffalo fish has shown that for fingerlings and larger fish, pellets provide a physical property which improves the economy of using artificial feeds.

Very small fish have shown better growth and feed conversion with pelleted feeds, even those which disintegrate before ingestion, than with the same formula in meal form.

The concentration of feed in a limited space and the inability of fish to select and reject certain ingredients, are factors in favour of using pellets. As fish grow larger, many species stop accepting small feed particles and suspended solids as supplied in meal-type feeds, and most of these "fines" are lost or become fertilizer. To be effective for some fish, feeds must retain a bite-size texture.

For fish with gill-rakers managed by a feeding programme of planktonic blooms and artificial supplemental feeds, pellets may be uneconomical if sufficient natural food is available.

However, in newly-filled ponds or those with new bottoms and in tanks, pools, and race ways, a complete feed in pellet form has resulted in better feed conversion than meal feeds. Once a natural food forage has been established in a pond and vitamin additives can be eliminated from the formula, the economics of pelleting versus meal feeding should be studied in terms of manufacturing cost and feed conversion.

It has been demonstrated that pellets can be used to selectively feed larger fish, permitting them to grow rapidly to market size without competing with smaller fish for the same feed.

Channel catfish (Ictalurus punctatus), blue catfish (Ictalurus catus) were observed over two growth seasons. Pellets of a standard formula were placed on saran screen trays of two types. One about 60 cm × 120 cm was lowered to the pond bottom and raised for loading and inspection by attachment to a tripod located on the levee (pond bank).

The other type was free to be placed directly on the pond bottom at any location and loaded and raised by hand. Pellets were also placed on the pond bottom and recovered by Ekman dredge. Fish were seined at intervals of ten minutes to an hour after feeding for analyses of stomach contents.

Table 1: Water Stability of Test Pellets made from a Standard Formula as Measured by the Percentage of Solids Retained on a Screen after Ten Minutes in Quiet Water

Treatment	*Percent Retained*
5% gelatinised corn solids (dry)	85.0
5% bentonite	88.6
Control formula - unground	90.0
5% guar meal	90.2
5% gelatinised corn solids - wet	92.0
Control formula ground through 2 mm screen	93.0
5% rice mill dust, 75 microns diametre	93.0
5% soybean flour	94.0
5% lignin sulfonate	94.0
5% rice mill dust sifted through 180 microns	96.2
10% rice mill dust sifted through 180 microns	98.2
20% rice mill dust sifted through 180 microns	98.5

Pellets of 3 mm diametre were readily accepted by fish weighing from 5 to 1 500 grams, all fish ingesting pellets of 3 to 9 mm length. Fish weighing more than 225 grams accepted pellets of 5 mm diametre × 5 mm length, and fish 454 grams and over accepted 6 mm diametre × 6 mm length pellets. Hard pellets of 13 mm diametre × 13 mm length were not readily accepted by any test fish.

Ten minutes after feeding in water at temperatures of 25 to 35° C, from five to nine percent of total fish weight was wet stomach contents. This was equivalent to one to three percent fish weight as dry pellets. In the process of ingesting pellets, fish consumed water so that stomach contents were found to be 70 to 80 percent moisture. One hour after feeding, about 75 percent of the total gastro-intestinal contents remained in the stomach, considerable amounts of protein and fat were gone from the food mass in the intestines, the water content in the stomach had increased, and the fish abdominal cavity was still greatly distended.

Hazards of Feeding Hard Pellets

For some fish there is a danger of over-feeding with hard pellets, causing swelling and rupture of the stomach. Over-feeding also mechanically forces pellets past the point of gastric activity, resulting in inefficient digestion. Another hazard with use of hard pellets may occur during winter feeding when fish ingest a full stomach capacity on a warm day followed by a period of cold weather. This may result in fermentation of undigested feed in the stomach with gas formation. Fish may float upside down, exposed to extremes in air temperatures and predation.

Pellet Hardness and Stability

Ideal pellets lure fish by their size, shape, and organoleptic properties and are available in water without loss of components. This last factor, called "pellet quality" is important for good feed conversion. A measure of pellet quality as applied to dry feeding conditions is the amount of small broken pieces of fines created during handling this may be related to the abrasive pressure required to break them down. However, dry pellet quality expressed as hardness or in the percentage of fines does not necessarily correlate with water stability. The average high-quality feed pellet made for poultry, swine, and beef, high in cereal grains, becomes mushy within minutes after immersion in water. There is evidence that coarse-textured feeds, water soluble, and hygroscopic ingredients weaken pellet structure in water, allowing components to separate and making the feed only partially accepted.

A laboratory technique for measuring pellet stability in water is useful for establishing feed formulation and processing specifications. One such test which has given reproducible values and shown good correlation with pellet recovery from feeding platforms and bottom sampling; devices, is as follows:

(a) 10 grams of pellets of equal diametre and length are distributed uniformly on a screen tray approximately 100 cm^2 in an area with raised sides. Mesh openings are slightly less than pellet diametre.

(b) Duplicate samples are lowered into quiet water of an aquarium or tank. If clumping or floating occurs, a piece of cut screen is placed on top of the pellets to hold them under water in a scattered position.

(c) After 10 minutes the trays are removed from water, slanted to drain, and placed in a moisture oven at 130°C for 2 hours, then cooled in a desiccator.

(d) The residue on the screen is recorded as dry solids not lost in water. Total solids in 10 grams of air-dry pellets are determined by oven-dry duplicate samples. Nutrients lost by leaching in water may be measured at this point by freeze-drying the residue on trays and comparing the chemical analysis with untreated pellets.

(e) The ratio of dry solids on the screen after 10 minutes in water to total dry solids in air-dry pellets, is used as a comparative measure of pellet water stability.

During the early days of using hard pellets for fish culture, a random selection of commercial feed was examined for water stability by this technique and none was found to have more than 60 percent retention after 10 minutes in water. To obtain more stable fish feed pellets and thereby improve feed conversion, guidelines for the industry have been proposed as follows:

(i) Before pelleting, grind the mixed feed through a 2 mm screen to an effective size of 125 microns.

(ii) Replace at least 5 percent of a non-essential ingredient with an organic flour such as rice dust, dried wood pulp liquor, wheat endosperm, or other binder.

(iii) Operate the pellet mill at its optimum rated amperage for maximum compression and extrusion pressures.

(iv) Add sufficient dry steam to condition the soft feed to a temperature of 85-90°C, thus causing gelatinisation of raw starch on the surface of all starch-containing ingredients.

(v) Cool and dry pellets before conveying to storage or bag-off bins.

It is typical of commercial catfish rations used in the southern part of the United States. Basically it complies with the requirements of a 30 percent protein feed, 25 percent of which is of animal source, with a protein to calorie ratio of 90 (grams protein in 100 grams feed × 100 divided by kcal/100 grams feed).

Table 2: Physical Measurements Of Pellet Stability of a Standard Catfish Feed

Process Variable	*Water Stability % Retained 10 minutes in Running Water*	*Broken by Stokes pellet Hardness Tester*
Unground, no steam, thin die	21.5	zero pressure
Unground, no steam, thick die	24.3	zero pressure
Unground, added steam, thin die	31.3	1 kg
Unground, added steam, thick die	78.9	3 kg

Ground, no steam, thin die	65.8	3 kg
Ground, no steam thick die	74.5	4 kg
Ground, added steam, thin die	84.9	8 kg
Ground, added steam, thick die	88.0	13 kg
Ground plus 20% gelatinised potato starch and 6% added water, no steam	98.9	20 kg

Note: Water movement through the test container may be used as an alternative to use of quiet water.

Formulae for common and mirror carp are similar to those for catfish with the exception that ground wheat will replace rice bran or wheat offal. High-wheat feeds are very water-stable if properly steam conditioned before compression and extrusion in a pellet mill.

Feeds low in, starch require a thick ring die and extra steam for conditioning. Special binders of starch origin provide some degree of water stability for pellets made with low-starch formulae.

Floating Pellets

Fish feeds may be pelleted by what is sometimes called the "extrusion" process, thereby expanding rather than compressing the various ingredients. Variations in formulation and processing allow a wide range in bulk density; 0.25. to 0.3 g/cc being common for fish feeding. This feature makes them attractive for certain types of fish culture. Fish may be observed while eating and the amount of feed limited to that which is accepted. The number of fish in a water impoundment and their health can be observed without sampling.

In general, the processing of expansion of extrusion consists of:

(a) conditioning soft feed which is in meal form to contain 25-30 percent water,

(b) conveying this conditioned feed by auger into a pressure-sealed cylinder,

(c) injecting steam, thus decreasing friction of material within the cylinder and increasing gelatinisation of raw starch,

(d) extruding to atmospheric pressure, almost exploding the material through holes in the die plate at the end of the cylinder,

(e) cutting off the extruding ribbon at the outside of the die plate by a rotating knife, and

(f) drying the pellets in a high-temperature oven at about 120 C to a moisture content suitable for storage.

Several details should be added to this general process description for a better understanding of floating pellets. The feed formula is important in obtaining a desired expanding texture. Cereal grains can be expanded to a very low bulk density, whereas protein concentrates low in starch may remain unaltered in bulk density.

Raw starch is a requirement of a good floating pellet; 90 percent being gelatinised during the 30 to 60 seconds the feed is in the expander cylinder. Pressure builds up to several atmospheres due to forward passage of the material into a smaller volume. Superimposed steam causes a high product temperature, changing the consistency of the material from a free-flowing meal to a dough. By the sudden release of pressure at the discharge end, the feed assumes a "puffed rice" texture like some breakfast foods and snacks.

Following oven drying, a standard pellet cooler is often used to lower product temperature after internal moisture is less than 13 percent. Even with this treatment, previous high temperatures partially destroy heat-labile vitamins and decrease the availability of some amino acids. Rather than over-fortifying the formula before pelleting, as is done for preparation of hard pellets, heat-sensitive additives may be sprayed onto expanded pellets after extrusion.

Biological tests under pond conditions with natural food available have shown no difference in growth of catfish and goldfish using a formula feed processed by hard pelleting and by the extrusion process. Also no significant difference was found chemically in total crude protein or in individual amino acids for several formula feeds each prepared as hard pellets and as extruded pellets. However, testing fish growth in tanks and cages has revealed a partial destruction of some nutrients in extruded feeds.

There is evidence that fish fed with floating pellets contain more liver and body fat than those fed the same formula feed processed as a hard pellet. This may be due to the increased digestibility of the carbohydrate part of the ration. The expansion process is expensive compared with other methods of feed manufacture in terms of equipment cost, heat used in generating steam and in oven-drying, and in loss of nutrients. It may be concluded that the value of extruded pellets-is best measured by practical feeding tests using data on feed conversion, cost of management, and acceptability by fish as deciding factors.

The Dravo Process for Non-Compacting Feed

In the early stages of development, fish are too small to eat anything other than finely ground particles. As the fish grow, it is necessary to feed larger particles containing a complete diet. Hence the development of pelleted feeds increases the availability of a balanced diet and at the same time reduces the tremendous loss that occurs when powdered or 'soupy' materials are fed to fish. The pellet size must be increased with the development of the fish.

The conventional method for manufacturing pellets involves forcing a feed mixture under pressure through a metal plate called a die. This die contains holes of the desired pellet size. As the mixture is extruded through the die holes, a rotating blade cuts the feed to the desired length. Although there are many variations in equipment design, all pelleting machines involve the extrusion principle. There are many disadvantages to this type of pelletizing operation. A large amount of pressure is required to force feed through the dies, which results in friction and, consequently, heat. Often pellets coming off the dies are so warm that they cannot be held in one's hand. The nutritive value of diets subjected to these conditions obviously is reduced through oxidation of heat-sensitive components. In addition, excessive heat and pressure frequently cause material to clog the dies, requiring stoppage of the process until the die plate is removed and cleaned. The operating cost is high, since the extruding operation requires considerable maintenance labour beyond that needed for production.

The dense, hard pelleted product from this operation sinks rapidly when placed in water. This minimises the opportunity for a fish to consume the pellet before it reaches the bottom of the pond or raceway. With those species. Of fish that will not eat off the bottom, the feed is lost. With those species that will eat off the bottom, the chances for contracting diseases are greatly increased.

A final disadvantage of the conventional pelletizing processes is that a limitation is placed upon the number of pellet sizes that may be produced. A 1/32 in (0.079 cm) pellet is about the smallest that can be manufactured by this technique, since it is most difficult to force these feeds through a smaller hole. All in all, these disadvantages can be overcome by a pelleting technique that produces a light, non-compacted pellet that will not sink rapidly in water and that has not been heated to the extent of destroying nutritional components in the product.

The Non-compacting Pellet

Process: It was discovered that dry fish feed forms a ball shape when sprayed with a water mist. on an oscillating table. The pellets were physically more desirable than those prepared by the more conventional means of manufacture. It was also discovered that this technique had been used in the fertilizer and chemical industries and that equipment was commercially available for production of the basic pellet.

A commercial Dravo pelletizing disc (originally developed for producing ceramic beads) is well suited for applying the non-compacting technique to making fish feed. Structurally, the Dravo pelletizer is simple. It consists of a disc that rotates at approximately a 45° angle. As the feed is tossed about on the disc, a fine spray mist causes the feed to form a ball shape of finely controlled particle sizes. Three metal projections aid in separating the particle sizes. The exact way in which pellets are produced can best be understood by using the number system that one finds on a clock and applying it to the rotating disc.

The operation is started by feeding a dry mixture onto the disc bed at approximately the six o'clock position. A fine spray of water is introduced at approximately the eleven o'clock position. Speed is first regulated so that the fine, dry mixture will be stopped about half-way up on the first disc projection. With continuing spraying and introduction of feed, pellets will form. The speed is then regulated so that the smaller pellets will collect about half-way up on the first disc projection. A flow pattern, as indicated by the arrows in the diagram, is achieved. Eventually uniform pellets will come off at the seven to eight o'clock position. This operation does not become efficient until a size concentration gradient of pellets builds up in the bed. The following may be done to decrease pellet size:

(i) increase the rate at which the mixture is fed into the disc; this reduces the time that the pellets may roll in the fines;

(ii) increase the disc angle;

(iii) feed closer to the three or four o'clock position; or

(iv) increase the speed of the disc.

Similarly, the following may be done to increase the size of the pellets:

(i) decrease the rate at which the dry mixture is fed into the disc;

(ii) decrease the disc angle;

(iii) feed closer to the 7 o'clock position; or

(iv) decrease the speed of the disc.

Using various combinations of controlling factors, any pellet size used in fish feeding may be obtained.

Pellets prepared by introducing the atomized droplets of water increases moisture content by 10-15 percent. Thus, pellets must be dried to remove this excess moisture. A conventional tray drier, utilising counter-current flow of warm air is normally used for drying the finished pellets.

Pelleted Product

A major advantage with the described type of pelleting technique is that much of the cost associated with maintenance is eliminated. There is no die to clog and remove for cleaning, so that the operation can run continually. The pelleting operation does not involve the heat or pressure associated with conventional methods of pellet manufacture. The retention of nutrients during pelleting is improved. The final product is soft, but hard enough to withstand normal handling procedures. The pellets usually float for a short period and then sink slowly. This gives the fish greater opportunity to consume the pellet before it reaches the bottom of a pond.

Another advantage is that many different sizes of pellets can be produced. As mentioned earlier, pellets are not made smaller than 1/32 in by conventional methods of manufacture. This means that the fish must go from a diet with flour-like consistency to a 1/32 in pellet or larger. This transition occurs at the most crucial, time in the animal's development, during the first weeks of life.

There are several disadvantages to overcome in using a pelletizing disc. Although a yield of over 90 percent is common, fines are produced as a result of pelleting and drying. It is necessary to recycle these fines for economic reasons. Further, differences in the specific weight of dietary ingredients can cause separation during the pelleting operation. This may be minimised by using raw material that is ground extra fine. In fact, fine uniform particle size is important to the efficiency of the pelletizing process.

Processing Plant for Producing Non-compacted Pellets

A generalised flow diagram of a non-compacting pelleting process is shown. There are several items of equipment necessary for this process and they are enumerated below.

(a) Meal storage bins, the size and number of the bins depend on the amount of material and the number of ingredients to be stored.

(b) A metring system for compounding the diet directly from storage bins.

(c) A ribbon blender, or other means of completely mixing compounded diet.

(d) Conveyors for moving the blended diet to the processing area.

(e) A storage tank for the blended meal; again, the size of this tank would depend upon the amount of meal being processed.

(f) Small tanks for storing additives such as vitamin premix and oil, to be blended just prior to pelletizing.

(g) Blenders for blending the final additives prior to pelletizing.

(h) A surge hopper for holding the blended feed.

(i) A metring feeder for supplying the diet to the pelletizing disc.

(j) A rotating, pelletizing disc such as that manufactured by the Dravo Corporation (Engineering Works Division, Neville Island, Pittsburgh, Pennsylvania 15225, USA).

(k) A conveyor and surge hopper to drier.

(l) A drier for reducing the water content from approximately 20-30 percent to below 10 percent. This drier should be so constructed as to cause minimum attrition of the damp particles.

(m) Screens for classifying pellets.

(n) A blower and lines for returning the screened fines to process.

(o) Bagging or packaging facilities.

Effects of Processing on the Nutritional Value of Feeds

Feeds are processed to facilitate handling and pelleting, and to increase feeding value by increasing digestibility or by inactivating specific growth inhibitors. Many processes that facilitate handling and pelleting increase the nutritional value of feeds as well, but the nutritional value of some feeds can be lowered by certain processes.

Heat Treatment

Heat treatment is employed to dry certain plant products and to improve the quality of pelleted feeds. Heat is sometimes employed in the extraction processes to remove oil from some oilseeds. Heat treatment improves the nutritional value of soybean meal by destroying the trypsin inhibitor which is naturally present, and by increasing the utilisation of proteins and amino acids, fats, and carbohydrates present in the meal. Digestibility is enhanced by partial "cooking" and thus the ME value is increased. Heat treatment increases the nutritional value of cereal grains by gelatinising starches and improving digestibility. This change occurs during steam pelleting, when dry steam is added to the mixture of feed ingredients just prior to pelleting to condition the feed so that better quality pellets are produced. Proper conditioning of feeds before pelleting results in improved pellet durability and a reduced amount of fines in the finished product.

Grinding

Grinding may increase the nutritional value of feeds by reducing the particle size of the ingredients and thereby increasing the surface area of the ingested feed and facilitating digestion. In addition, the grinding process adds metals to feeds from the grinding machinery and can prevent a micromineral deficiency. The metals that are added to feed are Fe, Zn, Cu, Mn, and Na.

Pelleting and Crumbling

The pelleting and crumbling process compacts the mixed feed ingredients and increases nutrient density and bulk density. For some species, pelleting improves palatability or acceptability of feeds. As mentioned above, the steam conditioning that occurs just before pelleting may improve digestibility. The heat generated during compaction of the pellet may also improve digestibility and destroy thermolabile toxic factors that naturally occur in some plant products. Pelleting enhances the availability of phosphorus in wheat bran and also permits the use of low density, bulky, unpalatable feeds that might not otherwise be practical to use.

Processing to Destroy Naturally Occurring Toxins and Inhibitors

Cottonseed Meal

Cottonseed meal contains gossypol, a compound with undesirable nutritional qualities. Free gossypol is associated with the pigment gland. Puncturing the pigment gland causes free gossypol to be mixed with other compounds in cottonseed meal and becomes bound. Bound gossypol is less harmful than is free gossypol.

The method used to remove the cottonseed oil affects the gossypol content of the residue which becomes cottonseed meal. Hydraulic pressed and solvent-extracted meals contain higher levels of gossypol than does prepressed solvent meal. Glandless meal has lower levels of gossypol. Treatment of the meal with iron salts eliminates the toxicity of gossypol, but also produces a darker meal which is unacceptable to the feed industry.

Treatment of cottonseed meal with phytase increases the availability of phosphorus, reduces gossypol toxicity, and increases the availability of some proteins.

Table 3: Toxins and Inhibitor Destruction by Processing

Feed Stuff	*Inhibitor*	*Deactivation Process*
Cottonseed Meal	Gossypol: Cyclopropane fatty acids; Phytate	Add iron salts; rupture pigment gland
Soybean Meal	Trypsin inhibitor	Heat, autoclaving
Linseed Meal	Crystalline water-soluble substance	Water treatment
Raw Fish	Thiaminase	Heat
Alfalfa Meal	Saponins; pectin methyl esterase	Limit amount fed
Rye	5-N-alkyl resorcinols	Limit amount fed
Sweet Clover	Dicoumarol	
Wheat Germ	Unidentified	Heat
Rapeseed	Isothiocyanate; thyroactive materials	

Soybean Meal

Heat destroys the trypsin inhibitor. Wet heat is more effective than dry heat.

Linseed Meal

The toxic factor for poultry can be removed by soaking the meal in water for 12-18 hours.

Effect of Processing on Moulds, Salmonella, and Other Harmful Substances

Moulds and Mycotoxins

Aflatoxins, produced on some feeds by Aspergillus flavus, are not inactivated by normal pelleting procedures. Prevention of mould growth is the best means to avoid contamination. Removal and disposal of mould-damaged material is essential. Certain materials that are sometimes contaminated by aflatoxins, such as corn, peanut meal, cottonseed meal, copra, and fish meal, should be monitored routinely to check for the presence of aflatoxins.

Salmonella

In contrast to aflatoxins, salmonella is killed by pelleting. Meat meal is a feed ingredient that should be routinely monitored for the presence of salmonella; proper sanitation procedures will reduce the incidence of contamination.

Effect of Processing on the Availability and Nutritional Value of Vitamins

Prolonged heat treatment can destroy the fat soluble vitamins and the water soluble vitamins thiamine, pantothenic acid, folic acid, and biotin. Such heat treatment can occur during the drying of protein supplements of plant and animal origin. For example, meat and bone meal contains little or no thiamine as a result of processing. The water soluble B vitamins are removed to some extent during the processing of fish into fish meal. The B vitamin content of fish-stick - water is higher than that of fish meal.

The pelleting process can alter the vitamin content of feeds. If insufficient anti-oxidants are present in the feed, pelleting may destroy Vitamins A, E, and K. On the other hand, pelleting may increase the availability of nicotinic acid and biotin, which are often present in feeds in a bound form.

Certain trace minerals act as pro-oxidants in feeds and can hasten the oxidative destruction of Vitamins A, D, and E. Manganese and iron are examples of such minerals. Non-nutritive feed additives, such as bentonite, which are added to feeds as binding agents, may interfere with the availability of riboflavin and certain divalent cations, such as zinc.

Some commercial methods for processing grains and oilseed meals result in decreased levels of certain vitamins in the final product. For example, coarse wheat flour grinders lose less Vitamin E than fine grinders. Germinated wheat may have a Vitamin E level as low as 30 percent of ungerminated wheat. Solvent extraction of soybean meal results in the removal of some of the Vitamins E and K and, thus, lower levels in the product.

Effect of Processing on the Availability and Nutritional Value of Trace Mineral Elements

Generally, the availability of minerals is not affected by processing. The availability of zinc is sometimes enhanced by addition of the chelate EDTA. Pelleting and crumbling do not affect trace mineral levels in feed.

Manufacturing Compounded Feeds in Developing Countries

Compound feeds for intensive fish culture have been developed in a few industrialised countries where preference for fanned fish is increasing. To meet animal protein needs in developing countries, efforts are being made to develop aquaculture on an intensive scale.

The major areas for improvement to increase productivity in aquaculture are synonymous to those in animal husbandry. These have been defined by FAO as:

(i) the improvement of stock through genetic selection, with the particular aim of breeding animals capable of high feed conversion,

(ii) the use of protective measures against diseases and pests,

(iii) the adoption of modern feeding practices using balanced rations, and

(iv) the use of feed additives to assist in achieving high feed conversion rates.

There are still many gaps in our knowledge concerning the nutrition of most fish species cultured in the tropics. Some progress has been made in other area; e.g., species selection, fish health, artificial spawning, and fish pond and hatchery management. This has made it possible for the successful practice of warmwater fish culture in a few developing countries.

Nevertheless, growth and productivity of animals (fish are not excepted) are keyed to the provision of nutritionally adequate diets. That feed cost in intensive fish culture represents the largest value input underscores the impact improved feed conversion will have on profitability in commercial aquaculture.

Research in warmwater fish nutrition continues to fill gaps concerning nutrient requirements for maximum growth and will undoubtedly lead to successful development of artificial diets based on tropical feed-stuffs.

Economic application of new technologies is geared to mass production of the new products. Where demand is sufficiently large, commercial production of formulated fish feed becomes feasible. In many instances, artificial pelleted fish feeds can be manufactured in mills designed for the production of feeds for livestock and poultry provided fine milling of properly selected ingredients permits the production of water stable pellets. The formulation of fish diets and the manufacturing processes involved have been discussed in earlier chapters and will not be reiterated here.

The principal cost in the manufacture of compound feeds is that of raw materials; this could amount to as much as 80 percent, or more, of the manufacturing costs in large size mills. Because of increasing transportation costs and the need to conserve foreign exchange, the tendency in most developing countries will be to use locally available ingredients. Although many developing countries remain food deficit, the supply of feedstuffs in some (especially of mill by-products and crop residues) is moving toward surplus as a result of the introduction of new high-yield crop varieties and expanded food processing practices.

Traditionally, the animal feed milling industry has been linked to the supply of raw materials. In recent years, due to increasing world demand, most feed ingredients have become articles of international commerce. As a result, quality has become more consistent for most feedstuffs. Availability is often just a matter of agreeing upon prices between trading partners. Feed mills, too, are now being located in areas where they can take best advantage of markets for finished products and of alternate supply sources for raw materials.

In countries where animal husbandry has not developed beyond small family holdings, expertise in commercial methods of compound feed manufacturing is lacking. Most developing countries now possess some commercial capability in animal feed production. In many of these countries, however, the feed processing industry is dominated by foreign enterprises which hold as propriety knowledge all vital information concerning the industry.

Compound feeds are unique among manufactured products in that they represent, as it were, a vital link in the human food chain. Like other links in natural food chains, one diet may differ from another generically, yet both satisfy to the same degree the nutrient requirements of the animal. For example, the dietary needs of rapidly growing broiler chickens can be just as satisfactorily met by a ration made from mixtures of tropical feeds tuffs such as rice bran, cassava, groundnut cake, copra cake, fish meal, and leaf meal, as by another ration made from mixtures of maize, soyabean meal, fish meal, and alfalfa meal.

Considering the urgent need for increased animal protein production and the under-utilisation of primary food resources in many developing countries, the establishment of viable feed processing industries in those countries represents a special challenge with broad socio-economic ramifications for national governments and almost unlimited opportunities for entrepreneurs.

For the reasons stated earlier concerning the uncommon flexibility of the feed industry, any attempt at describing details for the establishment of a "typical" feed mill in a particular geographic region will be too narrow in scope to serve the purpose of this training course. Instead, an attempt will be made to identify salient considerations that should precede the construction of such a "typical" mill.

Guidelines are also provided for data collection to enable the investor, public or private, to tailor the mill to particular needs.

Practical Fish Diets

Wet Diets for Research and Practical Feeding

Advantages of wet diets are:

(i) a pelleting machine is not needed, a food grinder will suffice;

(ii) many fish species find soft diets more palatable than dry, hard diets;

(iii) heating and drying is avoided which may prevent nutrient losses.

Disadvantages are:

(i) wet feeds are susceptible to micro-organism spoilage unless some preservation measure is followed;

(ii) oxygen sensitive nutrients such as ascorbic acid are subject to deterioration unless the wet diet is kept frozen; and

(iii) wet, non-heated animal tissues may contain an anti-thiamine enzyme.

Research Diets

Purified or semi-purified research diets are often processed as moist diets and stored frozen. In preparing laboratory research diets, the following specifications should be followed as completely as possible. All diets in an experiment should be:

(i) alike in all respects except the variable being tested;

(ii) palatable;

(iii) feedable (water stable, optimum particle size and texture);

(iv) nutritionally complete; and

(v) as far as possible made from purified ingredients.

Casein and gelatin (5:1) are a good protein combination for purified diets. Vitamin-free casein is available for use in vitamin experiments. Blood fibrin is a desirable protein for mineral studies. Egg protein is good for use in experiments where quantitative protein requirement is a variable. All of these protein sources are available in highly purified forms.

Dextrin is traditionally used as a carbohydrate source. Starch is satisfactory for warmwater fishes, but it is not utilised as well as dextrin by coldwater species. Some of the lipid should be from fish oil to provide omega-3 fatty acids; some may come from soybean oil or, for warmwater fish, animal fat.

Purified cellulose is used as a non-nutritive binder. Research has shown that up to 20 percent cellulose fibre is beneficial in purified fish diets, and levels above this are not deleterious provided the diet contains adequate amounts of the nutrients.

Moist diets must contain a binding agent that will hold the diet particle together for a reasonable time in water. Gelatin, agar, carboxymethylcellulose (CMC) and pregelatinised starch are good binding agents. Agar and CMC have no nutritional value and are used at levels of 2 to 5 percent.

Table 4: Test Diets for Fish

Ingredient	*Channel Catfish Diet*	*Test Diet H440 for Salmonids*
Vitamin-free casein	29	38
Gelatin	6	12
Dextrin	30	28

Cellulose flour	20.25	8
Fish oil	3	3
Soybean oil	3	-
Corn oil	-	6
Carboxymethylcellulose	3	-
Mineral mixture [1/]	4	4
Vitamin mixture [2/]	1.5	1
Calcium propionate	0.25	-

[1/] A salt mixture supplying all essential minerals for laboratory animals

[2/] Should provide the recommended vitamin allowances presented by NRC for warmwater fishes (1977) and for coldwater fishes (1973)

These diets have been used successfully with other fish species. Steps for preparing moist diets such as these are as follows:

(i) weigh each ingredient and place in separate containers; do not mix until everything is weighed for one diet;

(ii) mix the dry ingredients well before adding oil; then add oil and mix;

(iii) add approximately 350 ml of water per kg of diet mixture and stir with a dough mixer (Hobart); the moist mixture should have a stiff plastic consistency when compressed; if it does not stick together add more water; too much water will cause sticking when the extrusions come out of the food grinder;

(iv) extrude through a food grinder with proper diametre holes in the grinder plate; and

(v) break extrusions into short lengths by hand or with a sharp implement; package and freeze quickly.

Moist diets should be stored frozen until one or two days prior to feeding and should be kept in a refrigerator during this period in order to minimise loss of ascorbic acid.

Practical Wet Diets

In many areas pelleting equipment is unavailable, but the fish should receive their supplemental food in pellet form. Moist diets can be mixed and extruded through a food grinder into satisfactory pellets for feeding in water if a suitable binding agent is used. Natural feedstuffs that make good binders are fresh animal tissue and cooked starch. Blood, slaughter waste from fish, poultry or livestock, trash fish, and cooked grains or grain byproducts have been used successfully to bind moist fish feeds.

Wet, pelleted diets are made by mixing well-ground dry ingredients with ground wet materials and subsequently extruding the mixture through a food grinder with the optimum size holes in the grinder plate. Moisture content should be 35 to 40 percent; too much makes the extruded material too sticky to break up for feeding and too little reduces binding property. The combination of dry and wet ingredients depends on their availability and binding properties. For example, a combination of 50 percent trash fish and 50 percent rice bran will produce a diet mixture containing 63 percent dry matter and 23 percent protein (dry basis), with very good physical properties for feeding as wet pellets.

Wet feeds must be fed fresh or preserved against microbial spoilage. Some wet diets are frozen or refrigerated, but this is very expensive. Lactic fermentation and storage in air-tight containers is a satisfactory preservation method.

Feeding Lactic Fermentation Preserved Wet Diets to Catfish

A study was conducted at the Auburn University to evaluate lactic fermented catfish diets made to contain 60 percent catfish processing waste. The wet diets were stored in sealed containers for up to five months during the summer season. The wet diet mixture (40 percent moisture) decreased in pH to 4.5, due to lactic fermentation, within 48 hours, which prevented further bacterial decomposition. Exclusion of oxygen prevented mould growth (mould will grow at pH 4.5). The wet diet was removed from the container and extruded and broken up into pellets immediately before feeding.

Pour experimental diets were fed to channel catfish for a 5 month period. Diet 1 contained non-pasteurised fish waste, diet 2 contained pasteurised fish waste, diet 3 was similar to 1 and 2, but was dried and pelleted, and diet 4 contained marine fish meal instead of catfish waste and was in dry pelleted form.

Average percentage weight gain and food conversion for channel catfish fed the test diets were as follows:

Diet	*Percentage Gain*	*Food Conversion*
A Moist, nonheated waste	377	2.1
B Moist, heated waste	460	1.6
C Dry, catfish waste	462	1.5
D Dry, menhaden fish meal	624	1.1

The fish fed the diets containing the heated or dried catfish waste gained more than those fed the raw fish waste. Thiamine analysis of the diets at the time of feeding showed diet A to be almost devoid of this vitamin, which was assumed to be the reason for the poorer growth rate of the fish. Menhaden fish meal produced greater weight gain than catfish waste because of superior protein quality.

Table 5: Composition of Experimental Diets Containing Catfish Waste, Fed as Wet or Dry Pellets, and Fish Meal, Fed as a Dry Pellet

Ingredients	***Experimental Diets*** [c]			
	1	***2***	***3***	***4***
	%	%	%	%
Catfish waste	7.60.00	60.00	60.00	-
Commercial fish meal	-	-	-	15.17
Soybean meal	8.5	9.5	8.5	8.5
Peanut meal	8.5	8.5	8.5	8.5
Wheat flour	18.7	18.7	18.7	18.7
Lard	-	-	-	5.11
Dicalcium phosphate	-	-	-	1.04
Salt	2.0	2.0	2.0	2.0
Binder (carboxymethylcellulose)	2.0	2.0	2.0	2.0
Vitamin mix [a]	0.42	0.42	0.42	0.42
Water	-	-	-	38.56
Protein [b]	36.9	36.1	34.7	34.5
Ether/extract [b]	5.9	8.3	10.0	10.8
Ash [b]	9.1	13.7	13.1	11.9
Ca [b]	0.8	1.0	0.9	1.0
p [b]	1.0	1.3	1.2	1.2

[a] Vitamin mixture (mg/kg diet): vitamin A (30 000 I.U./g), 183.3; micro D, 73.7; alpha tocopherol, 119.9; ascorbic acid, 550; vitamin B_{12}, 41.6; riboflavin, 124.8; menadione sodium bisulfite, 13.7; calcium pantothenate, 187; niacin, 198; thiamine hydrochloride, 22; pyridoxine, 5.5; folic acid, 5.5; biotin, 0.44; choline chloride, 3 300; ethoxyquin, 198

[b] Average values as percentages of dry matter, determined by chemical analyses of the diets

[c] Diet 1 contained non-pasteurised catfish waste, fed wet,

Diet 2 contained pasteurised catfish waste, fed wet

Diet 3 contained pasteurised, dried catfish waste, fed dry

Diet 4 contained marine fish meal, fed dry;

A concurrent study was conducted to measure the rate of thiamine destruction in the moist fish food containing non-heated fish waste. After 12 hours storage at room temperature, the moist food contained 65 percent of the original thiamine level; after 7 days it contained 8.2 percent of the original level.

This indicates that moist diets containing non-heated fish waste can be fed with good results if fed within several hours after mixing. If fed after periods of storage, another source of thiamine must be provided for the fish.

Chapter 4

Catfish Feeds and Feeding

Catfish feeds represent the largest amount of commercially manufactured fish feed produced in the United States. The estimated 35 000 metric tons of farm-raised catfish produced in 1977 (Feedstuffs 1977) consumed approximately 60 000 tons of supplementary food. Almost all of these fish were produced in ponds and at least two-thirds of the feed fed was the extruded (floating) type. Feed conversion ratio in practice is near 1.7. Technology of formulating, manufacture, and feeding practical diets to catfish is advanced to the stage that highly productive and cost-efficient feeds are being produced; however, there are several areas which are still amenable to refinement.

Nutrient Requirements of Catfish

Protein and Amino Acids

The qualitative amino acid requirements of channel catfish have been determined. Catfish require the same ten essential amino acids as other finfish (salmonids, eel, carp). Quantitative requirements of essential amino acids have been partially determined for channel catfish and generally agree with values derived for other species. A notable difference is an apparent lower sulphur-amino acid-requirement for catfish than salmonids. The values which were not actually determined with catfish are estimated from values determined with other warmwater fish.

Table 1: Guidelines for Formulating Diets for Channel Catfish

Item	*Quantity in Diet*
Protein, %	30-36
Amino acids, % of the protein:	
Methionine + cystine	2.8
Lysine	5.1

Arginine [a/]	3.9
Tryptophan	0.5
Threonine	2.2
Valine	2.9
Histidine	1.6
Leucine	3.4
Isoleucine	2.3
Phenylalanine + tyrosine [b/]	5.1
Digestible energy, kcal/g of protein	7-8
Available phosphorus, %	0.5
Vitamins	c/
Trace minerals	d/

[a/] Argenine requirement determined for eel

[b/] Phenylalanine + tyrosine requirement determined for eel

The optimum percentage of protein in fish diets is influenced by several factors such as the following:

(i) Size of fish. Fish, like land animals, have higher protein requirements during early life than during later phases of growth.

(ii) Physiological function. Less protein is needed in a maintenance diet than in one fed for a rapid growth rate.

(iii) Protein quality. A protein that is deficient in one or more of the ten essential amino acids will produce less growth than a protein that is balanced in the essential amino acids, or, more of a low quality protein is needed in the diet for maximum growth than a high quality protein.

(iv) Non-protein energy in the diet. If the diet is deficient in energy, the fish will use part of the protein to meet energy needs thus reducing the amount of dietary protein available for growth.

(v) Feeding rate. Fish fed to less than satiation, as frequently occurs in intensive pond culture of food fish, will benefit from diets containing higher percentages of protein than fish diets fed at or near the satiation rate.

(vi) Natural foods. If natural aquatic organisms contribute significantly to the daily food intake of the fish, the protein level in the prepared diet may be reduced. For example, aquatic fauna that are consumed by various fishes contain from 60 to 80 percent protein, thus, if in abundance, the supplementary diet would need only a very low percentage of protein.

(vii) Economics. The cost and availability of protein sources is a major factor in determining how much protein to use in commercial diets.

Channel catfish of small size (10 to 20 g) require 9 to 10 g of high quality protein per kg of fish per day for maximum growth when fed a nutritionally balanced diet. As channel catfish approach harvestable size (0.5 kg) the daily protein requirement decreases to 7.5 to 8.0 g per kg. If channel catfish fingerlings will consume, or are fed, at the rate of 3 percent of their weight per day, a 33 percent protein ration will provide all of their protein requirement (10 g of protein per 30 g of feed per kg of fish). However, if fed at the rate of 2.5 percent of their weight per day, they would need a 40 percent protein diet. Most warmwater fishes have protein needs similar to channel catfish, thus protein levels of 30 to 36 percent will probably be adequate for most warmwater fish diets.

Energy Sources and Requirements

Energy requirements and-availability of energy from various feedstuffs for catfish are areas in need of more research information. Digestibility studies with channel catfish have indicated that they digest uncooked carbohydrate (starch) much better than salmonids but poorer than farm animals. Digestibility of calories in fats and in proteins is high. Cooking of feedstuffs, such as occurs in extrusion processing, improves the digestibility of calories in most materials, especially those high in starch. Fibrous materials, such as alfalfa meal and rice hulls., are poorly digested by catfish. With the exception of feedstuffs high in starch and highly fibrous, digestible energy values in livestock and poultry feeding tables are generally useful for catfish.

Energy needs of fish are less than those of warm-blooded animals because:

(i) fish do not have to maintain a constant body temperature;

(ii) fish require less energy for muscle activity to maintain their position in water than animals on land; and

(iii) fish require less energy to excrete nitrogen waste products than warm-blooded animals.

According to recommendations on the nutrient requirements of farm animals by the National Research Council, USA, the optimum amount of metabolisable energy for each gram of protein in the ration is 14 to 16 kcal for poultry and 15 to 24 for swine. This compares with values of 6 to 10 kcal per gram of protein which have been found adequate for catfish rations. Many commercial fish feeds, because of high protein percentages, are probably deficient in energy in relation to protein. The ratio of digestible energy to protein in several commercial catfish feeds was calculated and found to range from 6.6 to 7.4 kcal for each gram of protein.

Research at Mississippi State University, University of Georgia and Auburn University, has shown that the ration of energy to protein in channel catfish rations markedly affects growth. The values below represent the optimum ration of kcal of digestible energy per gram of protein for maximum weight gain by channel catfish between 30 and 500 grams in each study. Differences in the values may be due to variation in energy source or fish size.

Ratio of Digestible Energy (kcal) to Protein (g)	***Station***
9.6:1	Mississippi State University
8.3:1	University of Georgia
8.4:1	University of Georgia
7.9:1	Auburn University
7.3:1	Auburn University

Catfish eat to satisfy their metabolic energy requirement and, consequently, cease feeding when their calorie needs are satiated. Because of this phenomenon, fish will eat less of a high energy diet than a low energy diet. Therefore, too much energy in relation to the percentage of protein in the ration can prevent fish from consuming enough protein to meet their daily need for optimum growth rate, even though the fish are allowed to eat as much as they will consume.

High energy diets, especially when a high percentage of the calories is from fat, will produce fatty fish which have a lower dressing percentage than fish fed a lower energy diet. Experience with channel catfish has shown that body fat content only varies with very high energy diets if the non-protein calories in the ration are mostly from carbohydrates. High carbohydrate diets have not produced the adverse effects when fed to channel catfish that have been found with salmonids. In an experiment at Auburn University, when diets fed to channel catfish contained high percentages of protein (42 percent and above) and very low amounts of non-protein energy (less than 1.5 kcal per g), growth was suppressed.

When the protein level was reduced to 36 percent and the non-protein energy level remained the same, growth increased. When the non-protein energy in either the 42 percent or the 36 percent protein diets was increased, growth improved.

This experiment indicated that when too many of the calories in a fish diet came from protein, the efficiency of utilisation of the ration is suppressed, probably due to metabolisation of the protein for energy purposes by the fish.

Vitamins and Essential Fatty Acids

The vitamins B_{12} and inositol have been found to be unnecessary in diets otherwise nutritionally balanced. Qualitatively, the thirteen other vitamins have been demonstrated as dietary essential nutrients; however, the quantitative needs of all have not been elucidated.

Table 2: Recommended Amounts of Vitamins in Channel catfish Diets

Vitamin	*Amount mg/kg of Diet*
Vitamin A (I.U.)	5 000
Vitamin D_3 (I.U.)	1 000
Vitamin E (I.U.)	50
Vitamin K	10
Choline	550
Niocin	100
Riboflavin	20
Pyridoxine	20
Thiamin	20
D-calcium pantothenate	50
Biotin	0.1
Folacin	5
Ascorbic acid [a/]	50
Vitamin B_{12}	none
Inositol	none

[a/]Approximately 20 to 30 percent is lost in pelleting and 50 to 60 percent is lost in extrusion processing. Also, the half-life of ascorbic acid in fish feed is approximately 2.8 months in a warm environment

Essential fatty acid requirements for catfish have not been determined. Although trout require omega-3 fatty acids instead of omega-6, there is speculation among researchers that warmwater fish require both omega-3 and omega-6 fatty acids. Channel catfish grow poorly on fat-free diets; they grow only moderately well on corn oil diets, but well when soybean oil, fish oil and beef tallow is the lipid source. Practical diets containing 4 to 5 percent fat from natural feedstuffs have not produced effects indicative of a fatty acid deficiency.

Minerals

Calcium and phosphorus: Fish absorb large amounts of calcium from the water unless the calcium content of the environment is unusually low, i.e. less than around 5 ppm as $CaCO_3$. Common carp, rainbow trout, and red sea bream were found to absorb enough calcium from the environment to meet their needs for growth, provided their diets were adequate in phosphorus. Eel were found to need a dietary source of calcium. Channel catfish benefited only slightly from dietary calcium. The data were obtained in aquarium studies with a level of 35 ppm of $CaCO_3$ in the water. The graph indicates that the catfish showed a marked increase in growth response to the addition of phosphorus to the diet.

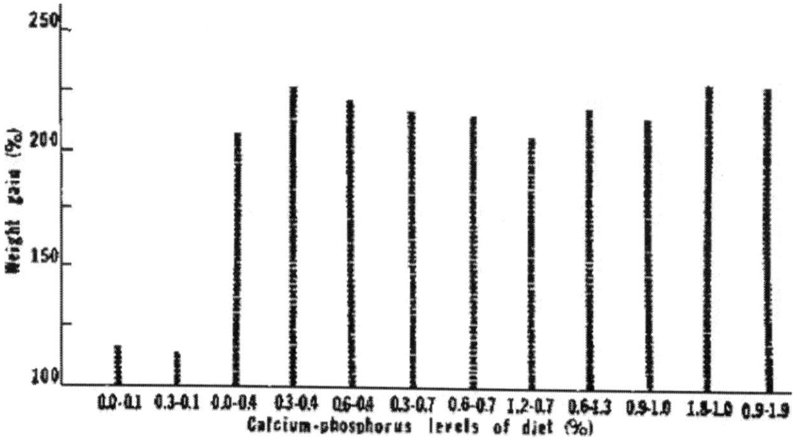

Figure 1: Weight gain of channel catfish fed calcium-deficient diet

Minimum requirements of "available" phosphorus in diets for channel catfish have been determined to be 0.4-0.5 percent using purified diets (Auburn University), and 0.8 percent using practical diet ingredients (University of Georgia). This difference is apparently associated with the availability of phosphorus to channel catfish from various dietary sources. If 0.5 percent is used as the minimum available phosphorus requirement for satisfactory growth of channel catfish, the following availability percentages are suggested for phosphorus from various common sources:

Source	*Availability of Phosphorus (percent)*
Dicalcium phosphate	8
Fish meal, meat, and bone meal	50
Soybean meal	40
Grains	33

Pond raised channel catfish fed an all-plant diet needed additional phosphorus in their ration for maximum growth rate, as indicated by the data in a feeding study at Auburn University. The amount of non-phytate phosphorus in the control diet was 0.22 percent. The addition of 0.3 percent phosphorus in the form of dicalcium phosphate increased the non-phytate phosphorus level to 0.55 percent which was adequate for maximum growth by the fish.

Other Minerals

Dietary requirements for most of other minerals have not been established for fish. Natural feedstuffs are usually adequate in K, Mg, Na, and Cl for normal growth of animals unless there is a high rate of mineral loss such as in sweating. These elements are probably available in sufficient quantity in practical fish feeds without mineral supplementation. However, fish feeds low in animal products (fish meal, meat and bone meal, etc.) may be deficient in trace minerals. When less than 15 percent of the ration is composed of animal products, a trace mineral supplement is recommended to provide the following amounts of minerals in the ration (mg/kg): Mn, 115; I, 2.8; Cu, 4.3; Zn, 88; Fe, 44; Co, 0.05.

Guidelines for Meeting Nutrient Requirements

Manufacture of Catfish Feeds:

Complete versus Supplemental Feeds: In many cases it is unnecessary and uneconomical to balance diets fed in ponds according to the absolute nutrient requirements of the fish. In this case, supplemental diets, which may be deficient in some vitamins or minerals, are often used to feed fish at low densities in the pond on the premise that the diet deficiencies will be made up by natural food organisms in the pond. In cases where the fish has limited or no access to other nutrient sources, such as in heavily stocked pond cultures or in artificial culture systems (cages, raceways and the like), all of the essential nutrients must be provided in the prepared diet in adequate quantities.

Such complete diets are being used more frequently in catfish culture ponds because fish density has been steadily increasing in recent years and dietary deficiencies such as the broken backs syndrome have been observed in pond culture.

These formulations have been used with good results in commercial culture.

Formulation Restrictions

Soybean meal and fish meal are used as major protein sources in all practical fish feeds in the United States. Soybean meal is the highest quality (best balanced in essential amino acids) plant protein available.

Fish meal is much more expensive than plant protein sources, but a minor amount of fish meal is necessary in all practical fish feeds. A minimum of 7.5 percent fish meal has been recommended in catfish feeds. Trout and salmon diets contain much higher percentages of fish meal; closer to 20 or 25 percent of the formula.

Cottonseed meal containing gossypol is toxic to some fish when fed in large amounts. Its toxicity has not been adequately evaluated with most fishes; therefore, when gossypol-containing cottonseed meal is used in fish diets, the level should probably not exceed 15 percent of the formula.

Table 3: Practical Diets Fed to Channel Catfish in Commercial Culture

Ingredient	*Skidaway, Ga. formula* [a/]	*Auburn No. 4 extruded diet formula* [b/]
	(%)	*(%)*
Fish meal (menhaden or anchovy)	10	10
Soybean meal (44% pro.)	35	51.2
Corn	29	22.7
Distillers dried solubles	-	7.5
Wheat	-	5
Corn gluten	20	-

Animal fat	2.5	2
Dicalcium phosphate	3	1
Vitamin premix	0.5	0.5
Trace mineral premix	0.05	0.08

[a/] 32 percent crude protein

[b/] 36 percent crude protein, 2.88 kcal digestible energy/g

Highly fibrous feeds tuffs like alfalfa meal, soybean mill-feed and rice hulls should be avoided since they are very poorly digested by fish. Fibre is beneficial for digestibility and absorption of nutrients in purified research diets, but studies with channel catfish indicated no benefit of fibre in practical diets.

Some ingredients affect the binding or pelleting property of fish feeds. Fat reduces the binding quality in both pelleted and extruded diets; consequently, supplemental fat should be applied after the feed has been polluted or extruded.

Usually, 2 percent or less of fat can be added to the ingredient mixture prior to processing without seriously reducing the binding property of the processed feed particles. Starch is an important ingredient in fish feeds for binding ingredients together. Generally, a minimum of 20 percent grain is necessary in extruded fish feeds for proper expansion during processing. Organic binding agents such as lignin sulphonates or hemicelluloses (both wood pulp byproducts) are often used in pelleted feeds, at a level of 2 to 2.5 percent, to increase water stability.

Pelleting and Extrusion

Pelleting, through compression, produces a dense pellet that sinks rapidly in water. Extrusion is a process through which the feed material is moistened, precooked, expanded, extruded and dried, producing a low-density feed particle which floats in water. Pelleting is less expensive and generally costs 10 to 12 percent less than extruded fish feeds. However, extruded or floating feeds are very popular with catfish farmers.

Pelleting involves the use of moisture, heat, and pressure to agglomerate ingredients into larger homogenous particles. Steam or hot water added to the ground feed mixture (mash) during pelleting gelatinises starch, which aids in binding ingredients. Generally, an amount of steam is added to the mash to increase its moisture content to approximately 16 percent and temperature to about 85 C before passing through the pellet die; however, ingredient composition will influence these conditions. The moisture must be removed by proper cooling and ventilation immediately after the pellets leave the pelleting apparatus.

Pellet quality refers to resistance to crumbling and water stability. The amounts of fat, fibre, or starch in the formula can influence quality of the pelleted feed. Some ingredients, because of chemical or physical properties, do not have desirable pelleting quality and can be used only in limited quantity in pelleted feeds.

Additives that serve primarily as pelleting aids are frequently used in fish feed formulas to reduce fines and increase water stability, although research in fish feed technology has demonstrated that high-quality fish feeds can be made without binding materials by following good pelleting procedures. However, use of compounds such as hemicellulose and cellulose derivatives, lignosulphonates, bentonites, and others does allow the processor greater variation in ingredient selection and processing conditions to produce pellets of satisfactory quality.

Extrusion requires higher levels of moisture, heat, and pressure than pelleting. Usually, the mixture of finely ground ingredients is conditioned with steam or water and may be precooked before entering the extruder. The mash, which contains around 25 percent moisture, is compacted and heated to 135 to 175 C under high pressure.

As the material is squeezed through die holes at the end of the extruder barrel, part of the water in the superheated dough immediately vapourizes and causes expansion. The low-density extruded particles contain more water than pellets and require more drying. Heat-sensitive vitamins are usually added topically after extrusion and drying. Extruded feeds are more firmly bound due to the almost complete gelatinisation of the starch and result in less fines than pellets.

Extruded or expanded fish feeds have two definite advantages over pelleted feeds: the particles float and are more resistant to disintegration in water, and a floating feed allows the fish culturist to observe the condition of the fish and the amount of food consumed. A large percentage of the catfish farmers in the United States use expanded feeds.

Catfish Feeding Practices

Pond Feeding: Most catfish produced in the United States are grown in ponds 10 to 40 acres (4 to 16.1 hectares) in size. The fish are usually fed once daily, six or seven days per week. Most catfish (approximately 67 percent or more) are fed extruded (floating) feeds. This allows the feeder to observe the feeding activity of the fish which prevents over-feeding and can serve as a check for disease or water quality problems. The fish are usually fed in the morning after dissolved oxygen (DO) level in the pond has begun to rise. (The typical diurnal variation in DO in highly enriched catfish ponds caused by the respiration and photosynthesis of phytoplankton).

Feeding should not be done late in the evening, so that maximum oxygen consumption of the fish will not coincide with a decrease in DO level in the ponds, as occurs when photosynthesis stops.

Water Temperature

The optimum temperature for food consumption and growth of channel catfish is near 86°F (36°C). Economical feeding can be done at temperatures above 70°F (21°C). The effect of water temperature on feeding activity of channel catfish in ponds during various parts of the season.

When minimum (morning) water temperature was above 26°C, feeding two times daily resulted in maximum food consumption and growth; when morning temperature was 22-26 °C, once per day feeding was optimal; and when morning water temperature was below 20°C, alternate day feeding resulted in the highest food consumption. These data indicate that catfish should be fed twice daily as soon as water temperature warms up in the early summer and fed on this schedule until the water temperature begins to decrease in the fall.

Winter Feeding

Several studies have been conducted on feeding regimes for catfish held in ponds to overwinter. Winter feeding can effect a small weight gain and will increase disease resistance of overwintered fish. Results of several studies at Auburn University indicate that feeding 1 percent of fish weight on days that water temperature is above 12°C is the best programme for fish 0.4-1.0 pound (454 g) size. Fingerlings may be fed at a rate of 1 percent of body weight, three times weekly throughout the winter.

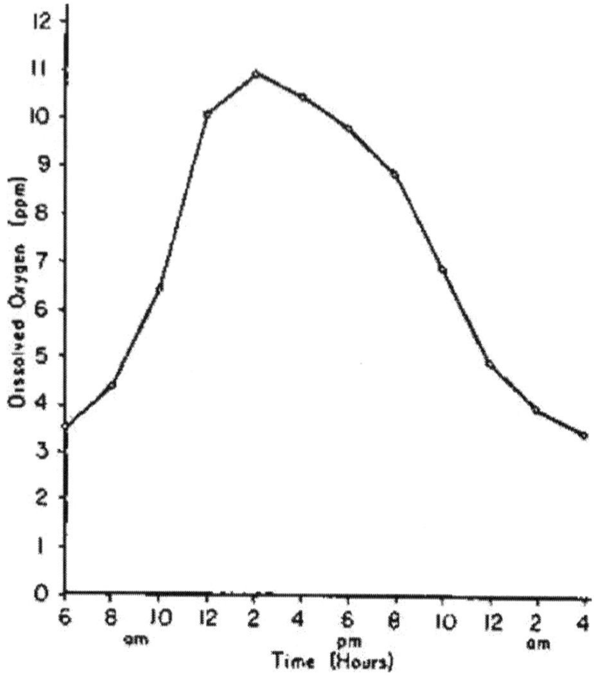

Figure 2: Typical daily variation in dissolved oxygen content in intensively fed channel catfish ponds

Feeding Rate in Ponds

Fingerling catfish will consume 4 percent or more of their weight in food daily, whereas catfish near 1-pound (454 g) size will consume only approximately 1.5 percent of the weight per day. Data in represent the daily amount of food consumed (percent of body weight) by channel catfish fed to satiation with two extruded (floating) diets.

The high energy-protein diet was less bulky than the lower energy-protein diet and the fish were able to consume more of it when appetite was greatest (during the early part of the season). The data in show the decrease in food consumption as fish size increases.

A study at Auburn University revealed that feeding catfish to 87 percent of satiation level produced as much growth and was more economical than feeding the fish to satiation when a 36 percent protein diet was fed, but not when a 30 percent protein diet was used.

Apparently, feeding at slightly below the satiation level and using a high protein diet is the most beneficial practice, since it results in more efficient food consumption.

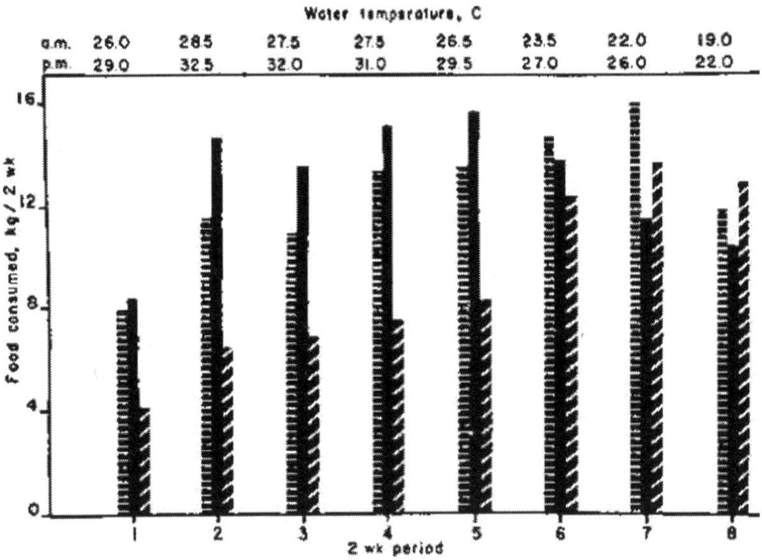

Figure 3: Food consumed per 2-week period by channel catfish according to three feeding schedules (Horizontal bars, once daily; solid line, twice daily; diagonal bars, alternate days) for 16 weeks in earthen ponds

Table 4: Weight (G) and Volume (Cm) of Food Consumed per 100 grams of Body Weight by Channel catfish Fed Diets of Two Nutrient Densities to Satiation

Date	***6/30-7/13***		***7/14-7/27***		***7/28-8/11***		***8/12-8/25***		***8/26-9/8***		***9/9-3/22***		***9/23-10/6***		***10/7-10/17***	
Temp . °C 1/	29.2		31.1		30.5		29.8		28.1		24.7		23.6		19.16	
Diet	g	cm^3	g	cm^3	g	cm^3	g	cm^3	g	cm^3	g	cm^3	g	cm^3	g	cm^3
High energy 2/	3.3	3.7	3.0	3.2	2.4	2.6	2.1	2.3	1.9	2.1	1.7	1.9	1.8	1.9	0.9	1.0
Low energy 2/	3.3	3.9	2.8-	3.4	2.1	2.5	2.1	2.6	2.0	2.4	2.0	2.4	2.0	2.4	1.3	1.5

1/ Average afternoon temperature at 1 metre depth

2/ Mass density of the diets was 0.91 g/cm^3 for the high energy diet and 0.83 g/cm^3 for the low energy diet

Crustacean Diets

Crustaceans, such as marine shrimps and freshwater prawns (Macrobrachium sp.), are becoming important species for commercial culture. The nutritional requirements of these animals appears to be similar to that of finfishes in many respects; however, some differences have been noted. For example, higher dietary levels of cholesterol are beneficial for crustaceans, and intestinal synthesis of amino acids appears to be significant. Because shrimp and prawns are slow feeders, diets that remain stable in water for several hours have been developed for feeding tests with these animals.

Pregelatinised starch (in extrusion cooked feeds), alginates, carboxymethylcellulose and other hydro-colloidal materials with good binding properties have been used in these diets. Extrusion-processed shrimp diets are commercially available.

In spite of the apparent need for water-stable diet particles for feeding crustaceans, a fair degree of success has been achieved in feeding freshwater prawns in ponds with pelleted poultry diets which have very poor water stability.

A study at Auburn showed that a pelleted diet, which was water stable for no longer than 15 to 20 minutes, produced almost the same weight gain in densely stocked freshwater prawn ponds as the same diet formula in extruded form, which had a stability time of 6-12 hours. It is speculated that the prawn receives a significant nutrient contribution from consuming microorganisms growing on the surface of the food particles of the disintegrated pellets in the pond.

In a recent study at Auburn University, a formula similar to that used for a practical catfish diet was processed in three forms for feeding freshwater prawns.

The diet mixture was extruded into high density (non-floating) cylindrical particles 3 mm diametre × 5 mm long and dried. Another mixture was extruded as described, but was not dried; it was coated with anti-caking and anti-mould compounds and fed as a soft diet. The diet was also fed in pelleted form. The dry and moist extruded forms have good water stability and the pelleted form disintegrated quickly in water.

The prawns were fed in plastic lined pools for 89 days and grew from an average size of 0.9 g to 6.4 g. There was no significant difference among weight gains of prawns fed the diet processed in the three forms. Prawns fed a commercial extruded prawn diet gained slightly but not significantly ($P>0.05$) more. Average food conversion for all treatments was 2.5 g weight gain/g food fed.

These results indicated that soft food particles or pellets with a high degree of water stability had no advantage over conventional pellets for prawns.

Table 5: Average Growth Rate and Food Conversion of Freshwater Prawns Fed a Diet Processed in Three Forms and a Commercial Diet

Diet [1/]	*Gain*		*Food conversion, g gain/g fed*
	Weight (g)	*Length (cm)*	
Extruded, dry	6.41	7.24	2.58
Extruded, moist	6.35	7.00	2.61
Pelleted	6.35	7.21	2.61
Commercial diet	6.97	7.56	2.37

[1/] Ingredient formula of the experimental feeds was: corn-wheat flakes, 47.6 percent; soybean meal (49 percent protein), 20.6 percent; anchovy fish meal, 18 percent; meat and bone meal, 4.6 percent; dried whey, 1.9 percent; animal fat, 6 percent; dicalcium phosphate, 0.7 percent; and a vitamin-trace mineral premix, 1.3 percent.

Novel Diets for Aquaculture

It is rather well agreed that a major problem in commercial aquaculture is the feeding of larval forms to, the juvenile stage. This particular problem is common to all species presently utilised and includes both fresh and saltwater teleost fish, crustaceans and mollusks. Many of these species are now the subject of intensive investigations to establish optimal conditions for profitable husbandry from egg to marketable size.

To date, culture of the more difficult to rear species has been dependent on live natural food such as artemia, algae, etc. However, mass culture of these natural live foods is both tedious and expensive; therefore it is necessary to develop synthetic or artificial diets (using natural readily available feedstuffs) to rear the larval forms.

Encapsulated Diets

It has been conjectured by researchers at Louisiana State University (LSU) that micro-encapsulation, in which liquids and particulate dietary components are enclosed within a carefully engineered wall with release under specific environmental conditions, has broad applications in fish and crustacean culture.

The technology associated with micro-encapsulation has not been described in scientific literature but has been the subject of several patents. Capsules consist of a wall and an internal phase. Release of the internal nutrients at active sites within the target biological system is a very interesting part of the capsulation technology. It is reported that a variety of mechanisms are possible to effect this release by modification of the wall material (composition, permeability, thickness, method of manufacture) during capsule production. Wall or shell dissolution can be accomplished by rupture (as in grinding), enzymatic action, pH change or bacterial action.

The wall could be a bio-degradable polymer i.e., modified gelatin, wherein the nutrients within the capsule could be released by the enzymatic processes of the animal or by micro-flora present in its gut. Ultimate selection of the particular wall matrix will depend on the material(s) to be encapsulated and specific properties desired in the finished product. It is reported that the range of components for the internal phase can constitute a complete diet for vertebrate/invertebrate culture. Capsules smaller than 20 microns in diametre are reported to be possible, allowing the culturist a wide range of sizes of nutritionally diverse capsules to satisfy the different growth stages of the particular species.

Advantages of encapsulated diets would be many. One major advantage is that specific nutritional requirements can be met with a high degree of precision since there will be minimal nutrient loss through water leaching.

Along the same lines, encapsulated rations would facilitate more rigid control over water quality, minimising BOD and adverse effects on oxygen, pH, etc. This is especially important under intensive culture conditions where good water quality is essential. Unlike natural or live foods which are by no means nutritionally complete, an encapsulated food could be thoroughly quality controlled, have good shelf-life, be consistent in nutrient composition, and may be developed completely free of contaminants.

The technology of fish feed encapsulation has not yet been demonstrated to be economic or technologically feasible - at least not in the scientific literature. The costs of micro-encapsulation could be considerable. However, Dr. S. P. Meyers of LSU believes that where survival and moulting rate in larval and young stages may dictate the success or failure of the aquacultural enterprise, great latitude is conceivable in cost projections of the support diet. Encapsulated systems might be far less costly and more dependable than currently used natural live food sources. Obviously, much research work is needed to ascertain the technological and economic feasibility of feed encapsulation.

Other Novel Forms of Fish Diets

Flakes and extruded diets should be also mentioned as other forms of processed feed that will afford reasonable water stability.

Flake diet formulation can be made with certain commercially available feedstuffs. The double drum dryer processing units are readily adapted to large-volume production. Vegetable binders can be used to achieve the water stability needed to optimise residence time of the flake in the aqueous system. Flakes can be reduced to smaller particle sizes without reducing the basic stability characteristics. Extrusion and extrusion-expansion of both intermediate-moist and dry type diets has also been suggested as a means of achieving some degree of water stability. Again, however, the most important consideration will be economic feasibility.

Microencapsulated Egg Diets for Fish Larvae

Egg is Unquestionably one of the most nutritionally balanced foods known for man and animals. Together with the facility with which it could be prepared and fed, and its general availability, egg was viewed with high expectations by early aquaculturists as a promising replacement for expensive live food such as Artemia in the feeding of very young fish. The ability of yolk and white to remain separate during cooking of unprocessed egg plus the fine granular texture of boiled egg yolk have indeed provided aquaculturists with a practical artificial diet that is superior to most other artificial feeds for this purpose.

Egg yolk per se is highly nutritious - perhaps more so than egg white. However, as a diet for very young fish its high energy-protein ratio could result in inadequate intake of protein necessary for maximum growth. At the same time, there are seemingly difficult problems concerning the preparation and feeding of whole egg to fish.

The ramifications of feeding egg to fish larvae will be explored in this chapter. Simple processing procedures for solving problems encountered in preparing a fish diet base on whole' egg will also be described.

Whole Egg versus Egg Yolk

Composition of Egg Components: It is clear that egg yolk is the more energy dense food although its protein content is 33 percent below that of whole egg. This disparity is exemplified by the very high energy-protein ratio of egg yolk. Animals, fish not excepted, eat to satisfy their energy requirements. A fish's food intake increases or decreases corresponding to a reduction or an increase in the diet's energy density. For diets of fixed protein composition, the requirement for this particular nutrient may not be sufficiently met if the diet has too high an energy-protein ratio.

Table 6: Composition of (chicken) Egg

	Whole egg	*egg white*	*egg yolk*
Protein, %	48.8	76.9	32.8
Fat, %	43.2	-	62.2
Gross energy, kcal/kg	5 830	3 070	6 910
Metabolisable energy (ME), kcal/kg	4 810	2 533	5 700
ME: protein ratio	9.8	3.3	17.3
Calcium, %	0.2063	0.0427	0.2653
Phosphorus, %	0.873	0.282	1.020
Amino Acids, %			
Arginine	2.968	4.179	2.369
Cystine	0.837	1.282	0.526
Isoleucine	2.734	4.307	1.896
Leucine	4.063	6.273	2.790
Lysine	3.047	4.427	2.369
Methionine	1.563	2.700	1.663
Phenylalanine	2.500	4.427	1.316
Threonine	2.500	3.692	1.843
Tryptophan	0.837	1.350	0.577
Tyrosine	1.952	3.076	1.316
Valine	3.674	6.025	2.263

Whole Egg versus Egg Yolk as Diet for Fish Larvae

The following calculations compare whole egg with egg yolk as an artificial diet for fish larvae.

Assume that a diet containing metabolisable energy (ME) at 4.0 kcal per kg represents a practical reference diet, with respect to energy density, for fish larvae.

(a) *Egg Yolk:* Amount of egg yolk required to provide the same amount of ME as 1 kg reference diet

$$= \frac{4.0}{5.7} 0.702 kg$$

0.702 kg egg yolk contains

0.702 × 0.328 = 0.230 kg protein

Therefore a diet consisting of 0.702 kg egg yolk plus 0.298 kg inert filler will have the same energy density as the reference diet and will contain 23.0 percent protein. This dietary protein level, we know, is inadequate for rapid growth of young fish.

(b) Whole Egg

Amount of whole egg required to provide the same amount of ME as 1 kg reference diet

$$= \frac{4.0}{4.81} = 0.832 kg$$

0.832 kg whole egg contains

0.832 × 0.488 = 0.406 kg protein

Therefore a diet consisting of 0.832 kg whole egg plus 0.168 kg inert filler will have the same energy density as the reference diet but will contain 40.6 percent protein.

Clearly, whole egg more adequately meets the protein level required in the diet of fish larvae.

Computations involving other nutrient parameters may similarly be made. Thus the whole egg diet will provide, per kg feed on an isocaloric basis, 15.47 mg and 2.30 mg of riboflavin and pantothenic acid, respectively, as opposed to 7.72 and 0.29, respectively by egg yolk.

Microencapsulated Diets Made from Whole Egg

Raw egg contains the growth inhibitor avidin which must be deactivated before the egg can be fed to fish. This is achieved by the application of heat. Cooking unprocessed egg, however, causes irreversible separation of the two egg fractions, yolk and white, due to denaturation of protein components in both fractions. Blending the two fractions, after cooking, info a homogenous water stable diet, requires the addition of suitable binders which add to cost.

The ideal product would be one that will possess the following characteristics:

(i) Wholesomeness - the product should be of high nutritive value and be free from the harmful effects of avidin

(ii) Consistency - size and texture of the product should favour ready acceptance by fish consuming the diet

(iii) Water stability - the product components, i.e. yolk and white, should not separate during feeding; feed particulates should have some buoyancy and remain in the water column for a sufficiently long period of time for the fish to eat it; the product should have minimum solubility in water but undergo some leaching of flavour to attract fish to the food

(iv) Low biological oxygen demand (BOD) - the product should not undergo rapid microbial degradation in water

(v) Good shelf life - the product should keep well under ordinary storage conditions

(vi) Ease of preparation - the product should be easy to prepare.

A hard boiled egg is a macroscopic representation of a product with such desirable properties. The egg eaten whole is highly nutritious and is free from avidin. The hardened albumin casing of hard boiled eggs prevents the separation of yolk. Heat denatured egg albumin is insoluble in water but some leaching of flavour occurs when the shelled, boiled egg is immersed in water for a period of time. Heat denatured protein is far more resistant to microbial action than is raw egg albumin. An intact boiled egg will stay 'fresh' for days under ambient conditions. Preparation of a hard boiled egg only involves the immersion of the unbroken raw egg in near boiling water for a short time period. In essence a hard boiled egg is a form of encapsulated food product in which the relatively unstable yolk is protected by a rather resistant albumin casing forming the capsule. If this can be brought down in scale, a similarly wholesome diet that is suitable for feeding fish larvae would be obtained.

Physical Properties of Eggs

When shelled eggs are placed in a glass container - taking special precaution not to damage the yolks - we have an intimate mixture of two substances, yolk and white.

The mixture is, in fact, a macro-colloidal system. The yolk constitutes the colloid, or dispersed phase, and, the white is the dispersing medium. This system also qualifies as an emulsion consisting of the stable dispersion of one liquid (fat) in a second immiscible liquid (water).

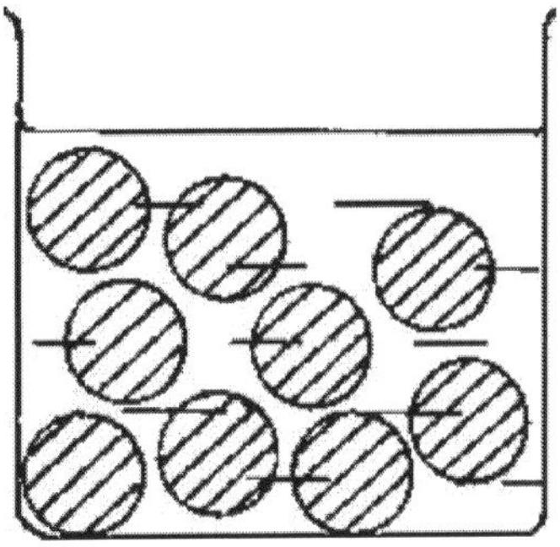

Figure 4: Unhomogenised whole eggs in container. Shaded spheres denote yolks.

Egg yolk is more than 60 percent fat on a dry weight basis. This fat is interspersed with phospholipids, nucleoprotein and lipoproteins which impart physical stability to the yolk. When boiled egg yolk is examined closely, it can be seen that it consists of aggregates of tiny spherical droplets of fat encased in heat coagulated protein capsules. This thin coating is insufficient to protect the lipid from oxidation by the atmosphere thus leading to rapid onset of rancidity and mould formation on the exposed product.

When raw whole egg is homogenised, the yolk is broken up and an ordinary emulsion is produced. This emulsion consists of countless microscopic-sized fat droplets (yolk) dispersed in the same immiscible watery medium (white). Coagulation of the dispersing medium will permanently coat the tiny lipid spheres to produce a microencapsulated whole egg product. The procedures for this process are simple and will be described in the following section.

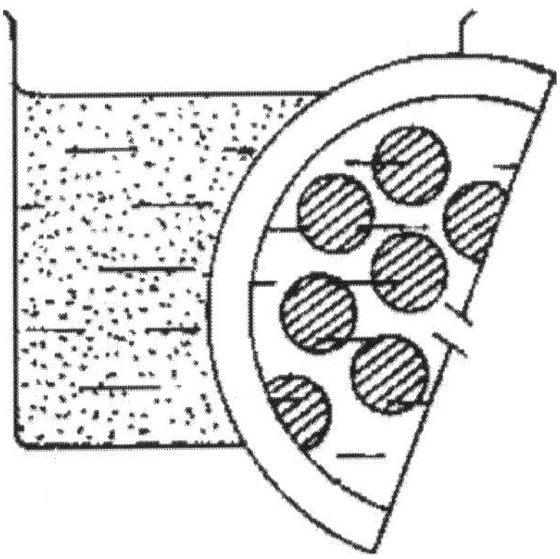

Figure 5: Homogenised whole eggs. Inset shows globules of egg yolk magnified.

Processing Techniques

Micro-encapsulated whole egg feed may be made by any of the following three methods.

Method A

Step	*Procedure*
1	Crack egg into heat resistant container.
2	Beat egg vigorously with fork or paddle; egg may also be homogenised with a mechanical blender.
3	Pour rapidly (approximately 150 cc for each egg) boiling water into homogenate, with constant stirring. A fine, opalescent suspension is obtained.
4	Make up to desired volume with cold water. A 50 g egg contains about 12 g dry matter.
5	Feed by the spoonful or scoopful directly to fish. The feed may also be applied as a spray using a 'knapsack' type sprayer.
6	Store unused feed in a tight container in a refrigerator.

Method B

Step	*Procedure*
1	Same as Method A
2	Same as Method A
3	Pour homogenate directly into boiling water. Amount of stirring depends on feed particle size desired. Initially strands of cooked egg are formed. Dispersal of egg yolk remains the same. Yolk is still encased in denatured (cooked) albumin.
4	Same as Method A
5	Feed resuspended material directly to fish. Do not apply with sprayer as this will result in feed particles of undesirably small size for large-size fry.
6	Same as Method A

Method C

Step	*Procedure*
1	Same as Method A
2	Add 50 cc water and proceed as in Method A.

3	Steam homogenate in same container or after transfer to appropriate dish or bowl. A custard made up of the solidified egg emulsion is formed.
4	Feed custard directly to fish after breaking up into desired size.
5	Store unused custard in tight container under refrigeration.

Advantages of the Novel Diets

Whole egg contains all the necessary nutrients needed during the first ten days of life of most species of fish. The diet prepared by Method A described in Section 3.2 possesses all the previously listed characteristics desired in feed for very young fish: i.e., for wholesomeness, consistency, water stability, low BOD, long shelf life, and ease of preparation. The opalescent protein coat of the micro capsule reflects enough light to attract fish to it. This also enables close monitoring of the feeding, process leading to reduction in feed wastage and maintenance of water quality.

The product from Method B is more suitable for larger fry and custard made with Method C is more suited to elvers which can be fed from a platform placed at the water surface.

Formulating Encapsulated Egg Diets

Shelled egg is lacking in the water soluble vitamins, especially ascorbic acid. An egg-based diet fed to fry for an extended period should be supplemented with vitamins and perhaps also calcium during Step 2 of the preparative procedure of each method. Similarly, a product of lower protein content may be obtained by adding high carbohydrate ingredients such as wheat flour or cassava flour. These ingredients must be finely ground, however.

The product from Method A can be freeze-dried, after removal of the water by centrifugation or filtration, to obtain an amorphous, dry, encapsulated feed.

Experimental Design in Diet Studies

Preparation of an experimental design before starting experiments greatly simplifies the collection and interpretation of data. Proper formulation of the hypothesis to be tested will help keep the experiment on course and will prevent the worker from trying to answer too many questions with one experiment.

Generally, an experiment is conducted to answer a specific question. A hypothesis is formulated and the experiment is conducted to test this hypothesis. The null hypothesis (H) is that there is no difference between a control treatment and an experimental treatment. If no statistical difference is found, the null hypothesis is accepted. If a statistical difference is found between treatments, the null hypothesis is rejected.

Experimental Designs

Four experimental designs are applicable to fish studies:

(i) The completely randomised design (CRD). The CRD is used when all of the experimental units are homogeneous.

(ii) The randomised complete block (RCB). The RCB is used when there is no homogeneity in one direction. The blocks are arranged to restore homogeneity among blocks.

(iii) The Latin square. The Latin square design is useful in situations where an experiment is conducted in a sectioned-off raceway where the experimental units should be shifted from one section to another periodically to randomise position effects.

(iv) The factorial design. The factorial design is used when treatments are combined and when an interaction between treatments is expected. The results can be analysed for significance between treatments (main effects) and also significant interactions can be tested.

Number of Observations (or Replicates) Needed in an Experiment

Treatments should be replicated and replicates should be randomised to obtain a valid error term; that is, one in which the variation within treatments is minimised by randomisation so that variation within each treatment is similar and from the same population of variation. The standard error of the mean (SEM) and the coefficient of variation (CV) are two common measures used in scientific literature.

One can calculate the number of observations needed in an experiment if one has an estimate of the variance (s^2) and knows the level of significance desired to show a significant difference. There are two ways to calculate this - a long way which has a confidence probability of 80-90 percent and a quick and shorter way which has a confidence probability of 50 percent. The long way is described below.

To determine the number of replicates, one needs to have an estimate of variance (s^2), a difference estimate (d^2 = difference between control and experimental treatments), confidence probability (t_0), and the level of significance probability (t_1).

$$r = \frac{2(t_0 + t_1)^2 s^2}{\delta^2}$$

If the expected difference between treatments is 2.5 percent in weight gain or amount of food eaten, and the variance (s) of this measurement in other experiments with the same species of fish is 1.31, the calculations would be as follows for an experiment with four treatments:

for r = 6,	$t_{0.2}$ with 25 df =	2.060	(from t table)
	$t_{0.05}$ with 25 df =	1.316	(from t table)
		3.176	

so

$$r = \frac{(2)(3.376)^2 1.31}{(2.5)^2} = 4.76$$

so

$$r = \frac{(2)(3.376)^2 1.31}{(2.5)^2} = 4.87$$

If there are five replicates, one can be 90 percent confident of picking up a difference of 2.5 percent, which would be a significant difference at the 5 percent level in a two-tail t test.

Testing for Significant Differences between Treatments

If two treatments are used, a simple t test is often the best method to employ. If three or more treatments are used, analysis of variance is the method of choice. In analysis of variance the variation in the experiment due to treatments can be separated from the variation due to experimental error, and thus significant differences between treatments can be detected.

Transformations

If data does not fit the model used in statistics, it can sometimes be transformed so that it does. Three common transformations are

(i) the square root;

(ii) logarithmic;

(iii) arc sine.

A discussion of transformations, experimental design and general statistics, can be found in a general textbook, like "Principles and Procedures of Statistics" by Steel and Torrie.

In order to truly answer hypotheses, most diet experiments must be designed such that all factors are the same between treatments except the one factor being studied. Ingredient testing, where, in a given formulation, one ingredient, say cottonseed meal, is replaced with another ingredient, say meat and bone meal, is a very slow and inefficient way to arrive at the "ideal" formulation. Changing ingredient prices makes an "ideal" formulation an ever-changing one anyway, unless cost is not a consideration. Diet experiments have basically one of two aims:

(i) to determine the requirements of fish for specific nutrients; or

(ii) to measure how well a particular feedstuff or feed formulation satisfies the nutritional requirements of fish or how available the nutrients of a particular feed are to fish.

Isocaloric and Isonitrogenous Diets

One common mistake in diet studies is the failure to insure that dietary treatments have the same caloric density (isocaloric) or the same protein level (isonitrogenous) when they should. For example, if an experiment is designed to measure the optimum protein level for growth of, say, 20-50 g coho salmon in saltwater pens, one approach might be to formulate practical diets that varied in protein level from 30-45 percent in 2.5 percent increments.

Protein level could be increased by adding herring meal to the diet and removing some basal feed, like wheat middlings, from the diet. The ME for herring meal is much higher (4 432 kcal/kg) than is that of wheat middlings (1 663 kcal/kg). Thus, simply exchanging ingredients will change the energy level of the diet as well as the protein level and confound the experiment with an extra variable. Energy must be added to some of the diets to make them isocaloric.

The same principle can be applied to experiments that include diets with different protein sources as dietary treatments. In this case, care must be taken to make sure that the growth effects being measured are due to protein source rather than protein level. The diets are made isonitrogenous by varying the level of the protein sources according to the protein content of each.

Control Diets

It is often difficult to compare the results of growth experiments done at different research stations or from year to year at the same station because of differences in stock of fish used and rearing conditions. This situation could be simplified by the use of standard control diets, such as the Oregon test diet or test diet H440 (Halver's diet). Researchers around the world have experience with these diets and know what kind of growth to expect from their use. On the other hand, not all experiments with test diets lend themselves to the use of appropriate reference diets.

Quality Control in Fish Feed Manufacturing

Quality control in the compound feed industry not only involves the verification of quality standards established for each feed ingredient as it is received into storage in the mill, but also involves the close monitoring of the quality of ingredients through the period of storage prior to usage and during its processing. Quality control continues as ingredients are merged during the mixing process and as they finally go into storage as compound feed.

Quality Control Procedures

Raw Materials: Feed manufacturing enterprises in developing countries operate under more difficult conditions than in developed countries. Problems of finance, management, and technical expertise aside, great difficulties in obtaining adequate supplies of raw materials are frequently encountered. Choices may also be restricted largely to feeds that are produced locally, since governments are reluctant to divert scarce foreign exchange for imports. Prices and availability of domestically produced feedstuffs are often subject to seasonal fluctuations while quality may also vary widely. Needed imports are made through agents with little experience in or responsibility for the quality of materials actually shipped. As a result, the need for quality control of both raw materials and finished products is greater for feed manufacturers in developing countries.

The purpose of quality control of raw materials is to ensure that minimum contract specifications are met. More precisely, it provides knowledge concerning the exact composition of raw materials and the level's of toxic substances normally present so that mixed feeds of the required nutritive value can be safely processed from them. These specifications are usually determined by a team consisting of the nutritionist, management personnel, and the quality control manager. The specifications relate to nutrient quality, cost, and the quality desired in the feedstuffs. Once the specifications are decided upon, they must be used and adhered to. Such a decision must be based on realistic reasoning, since it is useless to have specifications which are so narrow as to make purchasing impossible, or are so broad as to present severe problems for diet formulation.

Most contemporary feeds for livestock are formulated with the aid of a computer which calculates the cheapest product obtainable from available raw materials within constraints laid down for the nutritive value, toxicity, and palatability of the final product. Such least cost formulations are possible only when the composition of each raw material is known with a high degree of precision and when good quality control is maintained.

Preliminary Inspection

Materials received at the mill should be subjected to a thorough physical inspection to determine the following:

(a) evidence of wetting - mould growth confirms water damage;

(b) presence of scrap metals, stones, dirt, or other non-biological contaminants; and

(c) presence of insects.

Moisture content of feed grains should be determined by one of the rapid procedures commercially available. Any consignment exceeding 13 percent moisture is especially prone to insect and mould attack and should be separated from other stored ingredients. Preferably, materials with high moisture content should not be taken into the store until after drying.

Sampling

Sampling is the most important activity in quality control, because no analysis can be better than the sample from which it is made. Hence, proper procedures for taking representative samples for quality evaluation are essential.

Sampling of bagged ingredients is done with a spear probe. The probe is inserted diagonally and as horizontally as possible, from one corner of the bag to the other. In lots of 1-10 bags, all bags are sampled. In larger lots, 10 percent of all bags are sampled. Materials received in bulk are sampled by using a scoop, according to the size of the consignment. For smaller than 10 t consignments, two samples per ton are taken. Larger consignments, up to 100 t, require one sample per ton or one sample for every two tons depending on the size of the consignment.

Samples taken in the above manner should be pooled, thoroughly mixed, and then reduced in size by quartering to between 1 and 2 kg in weight.

Oil cakes and other coarse materials are sampled by random selection of pieces from different parts of the entire consignment. Five pieces per ton of materials are considered sufficient. The pieces should then be ground, mixed thoroughly, and the sample reduced in size to between 1 and 2 kg - as outlined above.

Samples submitted to the quality control laboratory should be placed in tightly sealed containers. Prior to chemical analysis, the samples must be reduced to a powder using a Waring type blender or a mortar.

Tests Required

For feedstuffs likely to contribute to both the protein and energy content of the final product, a "Weende" proximate analysis is usually conducted. This determines the moisture, crude protein, lipid, crude fibre, ash, and nitrogen-free extract content of the feedstuff. Methods of analysis are given in Section 4.

Additional tests should be carried out on materials having a high ash content to determine the proportion of acid insoluble ash present, thus providing an indication of the amount of sand or other dirt present. This will also enable detection of any deliberate contamination of expensive feedstuffs such as fishmeal. Salt (NaCl) analysis of fish meal is also required for these same reasons, as well as to avoid excessive levels of sodium in the final diet.

Calcium and phosphorus determinations are routinely conducted on mineral feeds, such as bone meal, calcium phosphate and calcium carbonate sources. Other feedstuffs such as fish meal, and meat and bone meal are also analysed for these two mineral elements.

Other Tests

The protein requirements of some cultured animal species are more exacting than others. For example, fish not only require higher protein levels in their diets but the protein quality requirements are also higher. The determination of amino acid content is a lengthy and complicated process requiring highly specialised, and expensive, equipment. However, it is advisable to test feedstuffs, such as fish meal, for lysine availability.

Certain feedstuffs contain natural toxins that, at high enough levels, are growth inhibitory and sometimes fatal to the animal consuming them. Principal among these are:

(a) *Urease:* An enzyme found in raw soybeans which produces toxicity through the hydrolysis of urea to toxic ammonia. The toasting process in soybean meal manufacture destroys the enzyme.

(b) *Gossypol:* An endogenous toxin present in the gland of cottonseed which persists during production of the meal unless removed by a special process, or, unless the cottonseed is a glandless variety.

(c) *Isothiocyanates:* Cyanogenic glycosides are found in linseed and cassava. Much of the toxicity is eliminated during processing of the raw material.

(d) *Aflatoxin:* Aflatoxin is a class of extremely potent toxins produced by the mould Aspergillus flavus. Aflatoxin may be present in any materials produced and stored under hot and humid conditions and is usually found in groundnut cake, palm kernel cake, copra cake, and maize (which have not been properly dried after harvest).

Molasses is a common feedstuff in the tropics. Periodic analysis is necessary to establish its sugar content. Molasses sometimes contain an undesirably high level of potassium and occasional checks on this should be conducted.

Fats are rapidly oxidised in the warm tropics. Rancidity causes off-flavour in feeds. Levels of free fatty acids in expeller cakes, such as palm oil and copra cakes, should be frequently checked. Unrefined palm oil or palm oil sludge, frequently used to increase the energy content of compounded feeds, contain high levels of free fatty acids.

Trace mineral and vitamin supplements form an important part of balanced diets for both fish and livestock. The availability of these feed components from reliable sources makes it quite unnecessary to perform routine checks on their quality unless large volume purchases are involved. Sending samples to a reliable commercial laboratory obviates the great expenses of equipping a laboratory for such testing.

Finished Products

Quality control of the finished products consists of determining the manufacturing process to ensure that ingredients were added in the proportions required by the formulation. Inhomogeneity of the final product, due to improper mixing or unwanted ingredient separation, thus, can be detected. If it can be established that raw materials conform with specifications and that process control is adequate, then only periodic checks on the finished products are necessary.

Preliminary Inspection

Most modern mills are equipped with sieves and magnets along the material flow lines for removal of tramp metal, rocks, and other scrap contaminants. However, smaller operations may not have such features, and physical inspection of the finished product should be carried out to determine the presence of such contaminants. Any detection of foreign contaminants should be brought to the attention of the mill supervisor who could then determine if the contaminants originated in the raw material or if they were the result of improper maintenance within the mill premises.

Sampling

To detect product inhomogeneity and significant ingredient separation during the manufacturing process, sampling should be obtained during bagging-off time by taking a handful from every fifth bag of 40-50 bags and pooling the individual samplings. Testing for variability is best conducted by probing the bottom, middle, and top of a bag with a short probe. Tests may be made on each sample, or the samples from the same bags may be combined and mixed before analysis depending upon the question to be answered.

The size of finished feed samples need only be half that required for raw materials, i.e., 0.5-1.0 kg after quartering.

Samples submitted for chemical analysis should be placed in tightly sealed containers.

Tests Required

It is not uncommon for feed to be despatched to the customer on the same day it is manufactured, a day or two before chemical analysis on the product is completed. Therefore, one senses the importance of proper raw material quality control and the apparent irrelevance of quality testing on finished products. Nonetheless, quality control of finished products is necessary because it serves two important functions:

(a) it checks the manufacturing process, and

(b) it checks the quality specifications, or claims, established for the finished products.

If raw material quality control is properly conducted and if process control is adequate, then the only chemical tests required of finished products on a regular basis are for moisture and crude protein. Periodic inspections on other components of the "Weende" proximate analysis should also be scheduled.

Raw Material and Finished Product Storage

Both raw materials and finished feeds undergo deteriorative changes during storage. For raw materials, these changes represent a direct economic loss because of the resultant decrease in nutritive value. Finished products undergo changes which not only lower their nutritive value below minimum specifications but also affect their palatability and appearance.

Due to seasonal fluctuations in availability and price, it is often necessary to maintain large inventories of raw materials if demands are to be continuously satisfied at stable price levels. Six-month inventories of stocks of imported raw materials are not uncommon. In the tropics, such lengthy storage periods present special problems, because the two environmental factors which most affect deteriorative changes in stored feeds are temperature and relative humidity. The rates of chemical changes that normally occur are increased with increasing ambient temperatures.

Insect and mould growth are also favoured by the warmer and more humid I climate of the tropics. In addition, there is a distinct relationship between insect activity and chemical changes in stored feedstuffs. A more thorough discussion of deteriorative changes in stored feed materials.

It suffices to state here that to minimise deterioration of stored raw materials and finished feeds, it is essential to keep them as cool and dry as possible) to minimise infestation by insect and rodent pests. Accordingly, warehouses should be constructed in such a way that the interior can be kept cool and dry with adequate ventilation to reduce micro-climatic variations created by the presence of the stored materials.

Methods of Analysis

Moisture

Weigh and place 4-5 g of the sample in a covered, flat, aluminium dish. Dry to constant weight at 100-105°C in a drying oven.

$$\text{Moisture content}(\%) = \frac{\text{Weight fresh sample} - \text{Weight dry sample}}{\text{Weight fresh sample}} \times 100.$$

Crude Protein (Kjeldahl Method)

Reagents:

(a) sulphuric acid (98%), nitrogen free,

(b) potassium sulphate, reagent grade,

(c) mercuric oxide, reagent grade,

(d) paraffin wax,

(e) sodium hydroxide, 40% solution,

(f) sodium sulphide, 4% solution,

(g) pumice chips,

(h) boric acid/indicator solution. Add 5 ml of indicator solution (0.1% methyl red and 0.2% bromocresol green in alcohol) to 1 litre saturated boric acid solution, and

(i) hydrochloric acid standard solution (0.1N).

Apparatus:

(a) macro Kjeldahl digestion and distillation units,

(b) Kjeldahl flasks (500 ml capacity or larger), and

(c) conical flasks, 250 ml.

Method: Accurately weigh 1 g of sample into a digestion- flask. Add 10 g potassium sulphate, 0.7 g mercuric oxide (pre-measured catalyst tablets containing these two reagents are available), and 20 ml sulphuric acid. Heat the flask gently at an inclined angle until frothing subsides and then boil until the solution clears. Continue boiling for an additional half hour. If the frothing is excessive, a small amount of paraffin wax may be added.

On cooling, add about 90 ml distilled water, recool, add 25 ml sulphide solution, and mix. Add a small piece of boiling chip to prevent bumping and 80 ml of sodium hydroxide solution while tilting the flask so that two layers are formed. Connect rapidly to the condenser unit, heat, and collect distilled ammonia in 50 ml boric acid/indicator solution. Collect 50 ml of distillate. On completion of distillation, remove the receiver (wash condenser tip) and titrate against standard acid solution.

Calculation: Nitrogen content of sample (%)

$$=\frac{(\text{ml acid}\times\text{normality of standard acid})}{\text{wt of sample(g)}}\times 0.014\times 100$$

Crude Fat

Reagents and equipment:

(a) petroleum ether (b.p. 40-60°C),

(b) extraction thimbles, and

(c) Soxhlet extraction apparatus.

Method: Weigh into an extraction thimble 2-3 g of the dried sample (residue from dry matter determination can be used). Place the thimble inside the Soxhlet apparatus. Connect a dry pre-weighed solvent flask beneath the apparatus and add the required quantity of solvent and connect to condenser. Adjust the heating rate to give a condensation rate of 2 to 3 drops/s and extract for 16 h. (The extraction time may be reduced to a minimum of six h by increasing the condensation rate.) On completion, remove the thimble and reclaim ether using the apparatus. Complete the removal of ether on a boiling bath and dry flask at 105°C for 30 min. Cool in a desiccator and weigh.

Calculation: Crude fat (% of DM)

$$=\frac{\text{weight of fat}}{\text{weight of sample}}\times\frac{100}{1}$$

Free Fatty Acids

Reagents and apparatus:

(a) ethyl alcohol,

(b) phenolphthalein (1% soln, in alcohol),

(c) sodium hydroxide (0.25N), and

(d) stoppered flasks, 250 ml. Method:

Weigh oil or fat into a stoppered flask and add 50 ml alcohol previously neutralised by adding sufficient 0.25N sodium hydroxide to give a faint pinkish colour with phenolphthalein (2 ml). Titrate with sodium hydroxide and vigorous shaking until a permanent faint pink colour appears.

Calculation: Free fatty acids % (as oleic acid)

$$=\frac{\text{g oil or fat}}{7.05}\times\text{volume of 0.25 N NaoH used in titration}$$

Retain extracted sample for crude fibre analysis and extracted fat for free fatty acid determinations.

Crude Fibre

Reagents:

(a) sulphuric acid solution (0.25 N),

(b) sodium hydroxide solution (0.313N),

(c) antifoam reagent (Octyl alcohol),

(d) ethyl alcohol, and

(e) hydrochloric acid, 1% v/v.

Apparatus:

(a) beakers, 600 ml tall-sided,

(b) round-bottom flask condenser unit,

(c) Buchner flasks, 1 litre,

(d) Buchner funnels. Hartley 3 section pattern,

(e) crucibles, silica with porous base, and

(f) rubber cones to fit above.

Method: Weigh about 2 g of the dried, fat-free sample into a 600 ml beaker. Add 200 ml of hot sulphuric acid, place the beaker under the condenser, and bring to, boiling within 1 min. Boil gently for exactly 30 min, using distilled water to maintain volume and to wash down particles adhering to the sides. Use antifoam if necessary. Filter through Whatman No. 541 paper in a Buchner funnel, using suction, and wash well with boiling water. Transfer residue back to beaker and add 200 ml hot sodium hydroxide solution. Replace under the condenser and again bring to boil within 1 min. After boiling for exactly 30 min, filter through porous crucible and wash with boiling water, 1% hydrochloric acid and then again with boiling water. Wash twice with alcohol, dry overnight at 100°C, cool, and weigh. Ash at 500°C for 3 h, cool, and weigh. Calculate the weight of fibre by difference.

Crude fibre (% of fat-free DM)

$$=\frac{\text{(weight crucible+dried residue)-(weight crucible+ashed residue)}}{\text{(weight of sample)}}\times 100$$

Ash

Weigh a 2 g sample into a dry, tared porcelain dish and then place in a muffle furnace at 600°C for 6 h. Cool in a desiccator and weigh.

Calculation:

Ash (%)

$$= \frac{\text{weight of ash}}{\text{weight of sample}} \times 100$$

Acid soluble and insoluble ash Reagents and apparatus:

(a) hydrochloric acid (1-2.5 v/v),

(b) filter paper, ashless, and

(c) dishes, porcelain.

Method: Use the residue obtained from the ash determination. Boil with 25 ml hydrochloric acid, taking care to avoid spattering, filter through ashless filter paper, and wash with hot water until acid free. Place filter paper and residue into a dry, tared porcelain dish and place in a muffle furnace at 600°C for 2 h or until carbon free.

Calculation:

Acid insoluble ash (%)

$$= \frac{\text{weight of acid-treated ash}}{\text{weight of sample}} \times 100$$

Calcium Reagents:

(a) hydrochloric acid (1-3 v/v),

(b) nitric acid (70%),

(c) ammonium hydroxide (1-1 v/v),

(d) methyl red indicator (dissolve 1 g in 200 ml alcohol),

(e) ammonium oxalate (4.2% solution),

(f) sulphuric acid (98%), and

(g) standard potassium permanganate solution (0.05 N).

Apparatus:

(a) porcelain dishes,

(b) volumetric flasks, 250 ml,

(c) beakers, 250 ml,

(d) quantitative filter paper and funnels, and

(e) burette.

Method: Weigh 2.5 g of finely ground material into a porcelain dish and ash as above (alternatively use residue from ash determination). Add 40 ml hydrochloric acid and a few drops of nitric acid to the residue, boil, cool, and transfer to a 250 ml volumetric flask. Dilute to volume and mix.

Pipette a suitable aliquot of the solution (100 ml for cereal feeds, 25 ml for mineral feeds) into a beaker, dilute to 100 ml and add 2 drops of methyl red. Add ammonium hydroxide one drop at a time until a brownish orange colour is obtained, then add two drops of hydrochloric acid to give a pink colour. Dilute with 50 ml water, boil, and add while stirring 10 ml of hot 4.2 percent ammonium oxalate solution. Adjust the pH with acid to bring back pink colour if necessary. Allow precipitate to settle out, and filter, washing precipitate with ammonium hydroxide solution (1.50 v/v). Place the filter paper with precipitate back in beaker and add a mixture of 125 ml water and 5 ml sulphuric acid. Heat to 70°C and titrate against the standard permanganate solution.

Calculation: Calcium (%)

$$= \frac{\text{ml permanganate solution}}{\text{wt sample}} \times \frac{\text{aliquot used(ml)}}{250} \times 0.1$$

Phosphorous Reagents

Molybdovanadate Reagent: Dissolve 40 g ammonium molybdate $4H_2O$ in 400 ml hot water and cool. Dissolve 2 g ammonium metavanadate in 250 ml hot water, cool, and add 450 ml 70 percent perchloric acid. Gradually add the molybdate solution to the vanadate solution with stirring and dilute to 2 litres.

Phosphorous Standards

Prepare stock solution by dissolving 8.788 g potassium dihydrogen orthophosphate in water and making up to 1 litre. Prepare the working solution by diluting the stock 1 in 20 (working concentrate 0.1 mg P/ml).

Apparatus:

(a) spectrophotometre to read at 400 mm, and

(b) graduated flasks, 100 ml.

Method: Pipette an aliquot of the sample solution prepared as for the calcium determination into a 100 ml flask and add 20 ml of the molybdovanadate reagent. Make up the volume, mix, and let stand for 10 min. Transfer aliquote of the working standard containing 0.5,

0.8, 1.0, and 1.5 mg phosphorus to 100 ml flasks and treat as above. Read sample at 400 my setting the 0.5 mg standard at 100 percent transmission. Determine mg phosphorus in each sample aliquot from a standard curve.

Sodium Chloride

Reagents:

(a) standard 0.1 N silver nitrate solution,

(b) standard 0.1 N ammonium thiocyanate solution,

(c) ferric indicator - saturated aqueous solution of ferric aluminium,

(d) potassium permanganate solution - 6% w/v,

(e) urea solution - 5% w/v, and

(f) acetone (A.R. grade).

Method:

Weigh 2 g sample into a 250 ml conical flask. Moisten sample with 20 ml water and then add, by pipette, 15 ml 0.1 N silver nitrate solution and mix well.

Add 20 ml concentrated nitric acid and 10 ml potassium permanganate solution and mix. Heat mixture continuously until liquid clears and nitrous fumes are evolved; then cool. Add 10 ml urea solution and allow to stand for 10 min. Add 10 ml acetone and 5 ml ferric indicator, and back titrate the excess silver nitrate with the 0.1 N thiocyanate solution to the red brown end point.

Calculation: Calculate results as sodium chloride,

$$\%\text{Nacl}=\frac{(15.00\text{-m}10.1\text{NNH}_4\text{CNS},\times 0.585)}{\text{g sample taken}}$$

Molasses Analysis

Total Sugars:

Reagents:

(a) Fehling's solution (Soxhlet modification)

(i) Dissolve 34 639 g of copper sulphate 5 H_2O in water and make up to 500 ml. Filter, and

(ii) Dissolve 173 g of Potassium sodium tartrate 4 H_2O and 50 g sodium hydroxide in water, dilute to 500 ml, stand for two days, and filter through prepared asbestos.

Invert Sugar Standards

Prepare stock solution by adding 5 ml of hydrochloric acid (sp.g 1.18) to 9.5 g of sucrose in solution and dilute to about 100 ml. After storing for two days at room temperature, dilute to 1 litre. Prepare working solutions (5 mg/ml) by pipetting 100 ml of the stock solution into a 200 ml volumetric flask, and neutralising with 20 percent sodium hydroxide using phenolphthalein as the indicator. Dilute to mark, and mix.

(a) hydrochloric acid (sp. g 1.18),

(b) hydrochloric acid (0.5 N),

(c) sodium hydroxide (20%),

(d) phenolphthalein indicator (1% solution in alcohol), and

(e) methylene blue indicator (1% aqueous solution).

Apparatus:

(a) electric heater, and

(b) conical flasks, 300 ml.

Method: Dissolve 8 g of liquid molasses and make up to 500 ml. Carry out an acid hydrolysis on 100 ml of the filtrate by adding 5 ml of hydrochloric acid (sp. g 1.18) and allow to stand for 24 h. Neutralise with sodium hydroxide (20 percent) using phenolphthalein as indicator, and then dilute to 200 ml.

Standardisation of Soxhlet solution. Pipette 10 ml of Soxhlet solutions (a) and (b) into a conical flask, mix, and add 30 ml of water. Add from a burette a volume of working standard that is almost sufficient to reduce the copper in the Soxhlet solution. Bring to boiling and continue boiling for two minutes. Add four drops of methylene blue and rapidly complete the titration, while still boiling, until a bright orange colour is resumed. Repeat several times and determine the volume of solution required to completely reduce 20 ml of the Soxhlet solution.

Titration of Sample: First, carry out an approximate titration: pipette 10 ml of solutions (a) and (b) into a flask and add 10 ml aliquot of-the sample solution. Add 40 ml of water and bring to boil. If blue colour persists, titrate with a standard working solution and calculate the approximate sugar content of the sample. To accurately determine the sugar content, pipette 10 ml of Soxhlet solutions (a) and (b) into a flask and add an aliquot of the sample solution. The volume of sample used will depend on the sugar content of the sample.

Table 7: Sample Volumes Used in Soxhlet Titration

ml H_2O	*ml sample*	*g sample in aliquot*	*Total sugar as invert, %*
40	10	0.08	73
35	15	0.12	82-58
30	20	0.16	61-41
25	25	0.20	49-35
20	30	0.24	41-29

(Reproduced from official Methods of Analysis of the AOAC, 1970)

Add water as indicated in the table, mix, and boil. During boiling, add a quantity of working standard from a burette so that the titration is nearly complete. Add methylene blue and complete the titration. Calculate the percentage sugar (as invert) by the formula:

% sugar = (F - M) × 1 × 100/W

where

F - is the volume of standard needed to reduce 20 ml of Soxhlet solution,

M - is the volume of standard sugar solution required to complete the back titration,

1 - is the weight of invert sugar in 1 ml of working standard, and

W - is the weight of sample in aliquot used.

Potassium

Reagents and equipment:

(a) hydrochloric acid (concentrated),

(b) potassium standard

To prepare stock solution (500 ppm K), dissolve 0.477 g potassium chloride (Analar) and make up to 500 ml with distilled water. To prepare working standard (10 ppm), dilute 1:50.

(c) silica crucibles,

(d) flame photometre, and

(e) muffle furnace.

Method: Dry 2 g of sample in a silica crucible at 100°C to expel moisture. Add a few drops of pure olive oil and heat over flame until swelling stops. Ash at 500°C in muffle furnace for 24 h, cool, and add 2 ml concentrated hydrochloric acid to dissolve the residue. Make up to 100 ml. Take 1 ml of this solution and make a further dilution to 100 ml.

Set the flame photometre to give a reading of 100 with the 10 ppm standard, and then read sample solution. If the sample reading does not fall between 50 and 100 make a fresh dilution to give an appropriate reading.

Anti-metabolite and Toxins in Feeds

Urease Activity in Soybean Meal:

Reagents:

(a) dimethylaminobenzaldehyde solution (DMAB): Dissolve 16 g DMAB in 1 litre 95% ethyl alcohol, and add 100 ml concentrated hydrochloric acid (stable for one month).

(b) pyrophosphate buffer: Dissolve 23.3 g $Na_4P_2O_7$ in approximately 980 ml distilled water. Add 3 ml of concentrated HCl and then additional HCl until the pH of the buffer is 7.7-7.8. Dilute to 1 litre.

(c) buffered urea solution: Dissolve 04 g urea in 1 litre pyrophosphate buffer (stable for 1 week).

(d) zinc acetate solution: Dissolve 22.0 g zinc acetate $2H_2O$ in distilled water, add 3 ml of glacial acetate acid, and dilute to 100 ml.

(e) potassium ferrocyanide solution: Dissolve 10.6 g $K_4Fe\ (CN)_6$, $3H_2O$ in distilled water, and dilute to 100 ml.

(f) charcoal.

Apparatus:

(a) water bath at 40°C, capable of maintaining temperature within ± 1°C, with shaking device,

(b) conical flasks, 125 ml,

(c) volumetric flasks, 25 ml, and

(d) spectrophotometre.

Method: Accurately weigh 1 g of soybean meal into a conical flask and add 50 ml of the buffered urea solution. Incubate in water bath for exactly 30 min at 40 C with shaking. Remove from water bath and quickly add 0.5 ml each of concentrated HCl, ferrocyanide solution, zinc acetate solution, and 0.1 g of charcoal. Shake for 15 min and filter. If the filtrate is coloured, repeat the procedure using more charcoal. Pipette 10 ml aliquots of the filtrate and the DMAB solution into a 25 ml volumetric flask and make up to volume with distilled water.

Make up also a reagent blank (10 ml DMAB made up to 25 ml with water) and a urea blank (10 ml buffered urea solution and 10 ml DMAB made up to 25 ml with water). Prepare a standard curve by pipetting aliquots of buffered urea solution from 2 to 12 ml into 25 ml volumetric flasks, adding 10 ml of DMAB and make up to volume.

Mix flasks well, stand in water bath at 25°C for 10 min, and then read at 430 mm. Calculate urease activity as mg/litre urea in urea blank less mg/litre urea in sample.

Free Gossypol in Cottonseed Meal

Free procedures are described for the determination of free gossypol: the first for normal meals, and the second for meals which have been chemically treated and contain dianilinogossypol.

Reagents:

(a) aqueous acetone, 7 parts acetone, 3 parts distilled water (v/v);

(b) Aqueous acetone - aniline solution

To 700 ml acetone and 300 ml distilled water add 0.5 ml redistilled aniline. Prepare solution daily.

(c) Aqueous isopropyl alcohol solution: 8 parts isopropyl alcohol, 2 parts distilled water (v/v),

(d) Aniline.

Distill reagent grade aniline over a small quantity of zinc dust, discarding the first and last 10 percent of the distillate. Store refrigerated in a brown glass stoppered bottle. Stable for several months.

(e) Standard gossypol solution

(i) Dissolve 25 mg of pure gossypol in aniline-free acetone and transfer to a 250 ml volumetric flask using 100 ml of acetone. Add 75 ml of distilled water, dilute to volume with acetone, and-mix.

(ii) Take 50 ml of solution (a), add 100 ml pure acetone, 60 ml of distilled water, mix, and dilute to 250 ml with pure acetone. Solution (b) contains 0.02 mg gossypol/ml and is stable for 24 h in darkness.

Apparatus:

(a) mechanical shaker,

(b) spectrophotometre,

(c) conical flasks, 250 ml capacity,

(d) volumetric flasks, 25 and 250 ml, and

(e) water bath (boiling).

Method: Grind sample to pass 1 mm screen taking care not to overheat. Take approximately 1 g of the sample and add 25 ml of pure acetone. Stir for a few minutes, filter, and divide filtrate into two. To one portion add a pellet of sodium hydroxide and heat in a water bath for a few minutes. A light yellow extract which does not change colour with sodium hydroxide indicates that the cottonseed meal is untreated and procedure (1) should be used. A deep orange red colour in the tube containing sodium hydroxide indicates the presence of dianilinogossypol and this requires that procedure (2) be used.

Procedure (1): weigh 0.5 to 1 g of sample, depending on expected gossypol content, into a conical flask and add glass beads. Pipette in 50 ml of aqueous acetone solution, stopper the flask, and shake for one hour. Filter, discarding the first few ml of filtrate, and then pipette out duplicate aliquots into 25 ml volumetric flasks. (Take aliquots from 2 to 10 ml, again depending on expected gossypol content.) Dilute one of the aliquots to volume with aqueous isopropyl alcohol (Solution a) while to the other aliquot (Solution b) add 2 ml redistilled aniline; heat in a boiling water bath for 30 min together with a reagent blank containing 2 ml of aniline and a volume of aqueous acetone solution equal to the sample aliquot. Remove solution b and the blank, add sufficient aqueous isopropyl alcohol to effect homogeneous solution, and cool to room temperature in a water bath. Dilute to volume with aqueous isopropyl alcohol.

Read samples at 400 mu. Set instrument to 0 absorbance with aqueous isopropyl alcohol, and determine absorbance of solution a and reagent blank. If the reagent blank is below 0.022 absorbance proceed as below, otherwise repeat the analysis using freshly-distilled aniline.

Determine the absorbance of solution b, with the reagent blank set at 0 absorbance. Calculate the corrected absorbance of the sample aliquot: the corrected absorbance is the absorbance of solution b minus the absorbance of solution a. Determine the mg of free gossypol present in the sample solution using the calibration curve.

Procedure (2): Weigh 1 g of sample into a conical flask, add 50 ml aqueous acetone, shake, and filter as above. Pipette duplicate aliquots of the filtrate (from 2 to 5 ml, depending on expected free gossypol level) into 25 ml volumetric flasks. Dilute one of the aliquots to volume (solution a) with aqueous isopropyl alcohol and leave for at least 30 min before reading on the spectrophotometre. Treat the other aliquot (solution b) as in procedure (1), determine the absorbances of solutions a and b as before, and calculate the apparent content of gossypol in both solutions a and b by using the calibration curve.

Preparation of calibration curve: pipette duplicate 1, 2, 3, 4, 5, 7, 8, and 10 aliquots of the 0.02 mg/ml gossypol standard into 25 ml volumetric flasks. Dilute one set (solution a) to volume with aqueous isopropyl alcohol and determine absorbances as previously. To the other set (solution b) add 2 ml of redistilled aniline and proceed as previously. Prepare one reagent blank, using 2 ml aniline and 10 ml of aqueous acetone, heated together with the standards. Determine absorbances as in procedure (1) and calculate the corrected optical density for each standard solution:

Corrected absorbance = (absorbance solution b - absorbance solution a). Plot the standard curve, plotting corrected absorbance against gossypol concentrate in the 25 ml volume.

Calculate free gossypol percent in normal meals as:

$$\text{Free gossypol\%} \frac{5G}{WV}$$

where

G - is the graph reading

W - sample weight

V - aliquot volume used

For chemically treated meals:

$$\text{Free gossypol\%} = \frac{5(B\text{-}A)}{WV}$$

where

A - mg apparent free gossypol in sample aliquot (a)

B - mg apparent free gossypol in sample aliquot (b)

W - sample weight

V - aliquot volume used

Thioglucoside Determination

The method described will give approximate thioglucoside content but does not allow the individual thioglucosides and isothiocyanates to be determined.

Reagents and apparatus:

(a) barium chloride (5% solution),

(b) volumetric flasks, 600 ml, and

(c) steam bath.

Method

To 10 g meal (de-fatted by Soxhlet extraction) add 250 ml distilled water, hydrolyse at 54°C for 1 h and then boil for 2h, keeping volume constant. Filter, retaining filtrate, and wash residue three times with 50 ml hot water. Add washings to initial filtrate and make up volume to 600 ml. Precipitate barium sulphate by heating and adding excess barium chloride solution. Leave on a steam bath for a few hours and then filter. Ash in a muffle furnace and then weigh precipitate.

Calculate approximate thioglucoside content as:

$$\%\text{trioglucoside} = \frac{(\text{M. wt. trioglucoside})(\text{Wt. of } BaSO_4)}{(\text{M. wt. } BaSO_4)(\text{Sample Wt.})}$$

Aflatoxin Analysis

A method of aflatoxin analysis is outlined below which is suitable for materials such as groundnut meal, coconut meal, and palm kernel meal. For full details of the method, and for alternative procedures reference should be made to Methods of Aflatoxin Analysis by B. D. Jones (1972), Report No. G70, Tropical Products Institute, London.

Reagents:

(a) chloroform (reagent grade),

(b) diethyl ether (reagent grade),

(c) chloroform/methanol mixture (95/5 v/v),

(d) "Celite", diatomaceous earth,

(e) Kieselgel 'G' (Merck),

(f) qualitative standard.

Helps to distinguish aflatoxin spots from other fluorescent spots which may be present. A groundnut meal containing aflatoxins B, obtainable from the Tropical Products Institute, London, can be used for this purpose.

Apparatus:

(a) thin layer chromatography plates, 20 X 20 cm,

(b) UV lamp, peak emission at 365,

(c) bottles, wide-mouthed, 250 ml,

(d) micropipettes, and

(e) shaking device.

Method

Weigh 10 ml of material into a wide mouthed bottle and thoroughly mix in 10 ml of water. (If high fat material is used, a prior Soxhlet extraction with petroleum ether will be necessary.) Add 100 ml of chloroform, stopper with a chloroform resistant bung, and shake for 30 min. Filter the extract through "Celite", take 20 ml of filtrate and make up to 25 ml (solution a). Take another 20 ml of filtrate and concentrate to 5 ml (solution b).

Prepare thin layer plates by shaking Kieselgel 'G' (100 g) with water (220 ml) for 20 min and apply the mixture to the plates with a suitable apparatus to a depth of 509 m. Leave for 1 h, then dry at 100°C. Spot 10 and 20 m of solution b, and 5 and 10 m of solution a onto a plate, together with a qualitative standard spot, in a line 2 cm from the bottom of the plate and at least 2 cm in from each side. Carry out the spot application in subdued light.

Develop the plate in diethyl ether to a height of 12 cm. Allow to dry in subdued light then redevelop the plate in chloroform methanol (95/5, v/v) to a height of 10 cm from the baseline. Examine the plate in a dark room, 30 cm from the UV source. The presence of a blue fluorescent spot at Rf 0.5 to 0.55 indicates aflatoxin B (check that the standard spot also lies in this range). The presence of a second spot at Rf 0.45 to 5 indicates aflatoxin G. The toxicity level of a sample can then be classified in terms of aflatoxins B and G.

Table 8: Toxicity Levels for Alatoxins B and G

Vol. applied (μ l)	*Concentration of aflatoxins (μ g/kg)*		*Toxicity level of fluorescence observed*
	No fluorescence	*With fluorescence*	
5μ (soln. a)	< 1000	> 1000	very high
10μ (soln. a)	< 500	500-1000	high
10μ (soln. b)	< 100	100-500	medium
20μ (soln. b)	< 50	50-100	low

The Art (Science) of Pelleting

Pelleting has been, and continues to be, a popular processing technique in feed manufacturing. In basic terms, pelleting converts a finely ground blend of ingredients into dense, free flowing agglomerates (pellets). There are many reasons used to justify the process, but it may be appropriate to list just a few:

- improved animal performance,
- decreased feed wastage,
- reduced selective feeding,
- improved bulk density,
- better material handling characteristics,
- destruction of deleterious organisms, and
- customer expectations.

Pelleting operations are not without cost. It is a fairly expensive process in terms of both capital and variable costs, but the expense is usually justified in improved plant profit as well as animal performance.

The purpose of this paper is to discuss the pelleting process in terms of operations and to describe how the success or failure of the operation can impact on profits as well as animal performance.

The Process

The formation of the pellet actually occurs at the "nip" between the rolls and the die. All other activities associated with the operation such as conditioning, cooling, etc. really support and augment the action at that point in the system. In order to understand the process and be in a position to make intelligent decision to improve throughput, quality or appearance, one must have a thorough understanding of what happens at the nip point.

Depending upon the physical characteristics of the feed, a lesser or greater proportion of the work done by the pellet mill is used for compression. For example, if the formula contains a high level of fibrous ingredients such as bagasse, bran, or ground alfalfa, the mill will expend a large amount of energy simply compressing the mash to the density of the subsequent pellet. Conversely, for a relatively dense feed such as high grain and soy meal, the mill will expend a lesser amount of energy for compression and a greater amount for throughput.

The "extrusion area" shown in the point at which the mash has reached pellet density and begins to flow through the die holes. There are many physical forces that must be dealt with in the pelleting process.

The primary purpose of the roll is to provide a force on the mash to density the feed and cause it to flow toward the die. The gap between the roll and the die, the roll surface characteristics and the physical properties of the mash determine how great this potential force might be.

The die provides, not only the final diametre of the pellet, but the resistance force on the feed and has a direct influence on throughput rate and pellet quality. These two forces (roll and die) are apposite each other as shown in the diagram, but must work together to provide quality pellets at an acceptable production rate. The force generated by the roll must be greater than the resistive force provided by the die; if not, throughput is zero.

With a general understanding of the process inside the pellet chamber, it is appropriate to move to a discussion of various factors that affect both throughput and pellet quality.

Pellet Quality

For purposes of this paper, pellet quality will be equated to the ability of pellets to withstand repeated handling without excessive breakage or fines generation. There are many factors that affect pellet quality, but the following will be discussed in some detail:

- formulation
- ingredient particle size
- mash conditioning
- feed rate

- die speed
- die specifications (design)
- other factors

Formulation

There are feedstuff materials that pellet well and produce a durable pellet and others that will not. MacBain (1966) developed a pelletability chart in which he ranked feed ingredients in their pelletability and degree or abrasiveness.

Bartikoski (1962) experimented with applying numerical value to each major (feed) ingredient to indicate its "stickiness" or its ability to help form a tough, durable pellet. He called that value a "stick factor" and fed that factor into the computer along with the various nutritive values of each ingredient to provide formulas that meet all nutritional specifications as well as supplying a formula that will produce a quality pellet at least cost.

Those early workers led others to experiment with the effects of various ingredients - grains, milled grain by-products, fats, pellet binders, minerals, etc. - on pellet quality or durability. They also led to the development of a standard method for testing pellet durability perfected in the 1960's by Dr. Harry B. Pfost at Kansas State University and accepted as a standard by the American Association of Agricultural Engineer - ASAE S-269.3 (ASAE, 2003).

That method is generally known as the K-State, or tumbling can, durability test; and it provided a means of quantifying the toughness of pellets or their ability to withstand the downstream handling that is typical in feed plants and feed delivery systems. That was a major breakthrough in the technology of pelleting and has served the industry for all these years.

Minerals

Pellet mill performance can be significantly affected by the physical and chemical forms of the calcium and phosphorus sources used in the formula. Sutton (1979) investigated the effect of deflourinated phosphate (two particle sizes) and dicalcium phosphate (18.5%) on pellet mill performance with a broiler grower formula. He found the production rate for the diet containing regular grind deflourinated phosphate to be 68.9% greater than for the diet containing an equal amount of dicalcium phosphate. The finely ground deflourinated phosphate had a 52.5% advantage over dicalcium phosphate. In a similar study (Behnke, 1981) we also studied the effect of mineral sources on pellet mill performance and pellet quality.

Two deflourinated phosphate sources, a fine grind (DPF) and a regular grind (DPR), and an 18.5% dicalcium phosphate (DCP) were used. A practical layer diet was used in which each test mineral source was evaluated at both high (2.5%) and low (1.5%) levels in the diet. At both levels tested, the production rate for the deflourinated phosphate sources significantly outperformed dicalcium phosphate; while the DCP had a slightly, but not significantly, higher pellet durability index. That would indicate that a physical change - thicker die or reduced feed rate - could be made to improve pellet quality without a substantial loss of system throughput. Behnke (ibid), Verner (1988), and McEllhiney and Zarr (1983) reported similar results comparing phosphorus sources in a variety of pelleted feeds produced under many conditions.

Those studies are cited, not to encourage or discourage the use of any mineral source or any other ingredient - that's the nutritionist's decision - but to indicate that those sources and ingredients can affect pellet quality and production rate and should be considered in the quest for improved pellet quality.

Particle Size

Optimum particle size for best pelleting results has been a matter of controversy for almost as long as feeds have been pelleted. Young (1960) found no significant differences in pellet durability when he experimented with feed rations containing 40, 60, and 70 percent ground corn or grain sorghum when the grain portions were ground coarse, medium, and fine.

Martin (1984) compared pelleting efficiencies and durabilities using a hammermill and a roller mill to grind the corn portion (59.5%) of a pelleted feed. He did not find any differences ($P<.05$) among the various treatments. The average particle size of the hammer milled corn (3.2 mm and 6.4mm screens) ranged from 595 to 876 microns, and the roller milled corn (fine and coarse) ranged from 916 to 1460 microns. Stevens (1987) conducted similar experiments in which No. 2 yellow corn was used as the grain portion of the typical swine formula. The corn was ground with a hammermill through three screen sizes: 1/16"(1.6 mm) (fine); 1/8"(3.2 mm) (medium) and 1/4"(6.4

mm) (coarse). He then measured the effect of the ground grain particle size on the pelleting production rate, electrical efficiency, and pellet durability. There were no significant ($P<.05$) differences in the pelleting production rate or PDI values from different particle sizes of corn mixed into the swine ration; although, the total electricity required to grind the corn and pellet the mash was significantly greater for the fine ground corn. When ground wheat was used as the grain portion of the swine ration, pellet production rates and PDIs improved as the grain was ground finer; but the finer ground wheat also required substantially more electrical energy.

While the research cited may seem to provide conflicting results, there is overwhelming evidence that the average particle size of the ground grain portion of a ration or of the total ration (mash) affects the pelleting process - throughput and/or pellet quality. The effects, simply, are not the same under all conditions or for all rations. The operators must conduct their own research under their own operating conditions and on the feeds that they produce.

It is common that some portion of a plant's product mix is often in mash or meal form and that grinding the grains more finely in a pre-grind system or the whole mix in a post-grind system causes handling problems in those mash feeds. There are two solutions to that dilemma - either provide two ground grain bins over the mixing system or find a grind (particle size) in the middle somewhere that will produce the better quality pellet and still provide the flowability or angle of repose that is needed for mash feeds. The first option is, of course, the better one but may not be possible, or too expensive, in a given grinding/mixing system situation.

Grind as fine as you must for best pellet quality in your operation and with your operation and with your feed rations, but don't over grind. That is wasteful of energy, reduces production rates, adds to manufacturing costs, and may do more harm than good to the consuming animal.

Mash Conditioning

This is a subject unto itself and will not be addressed in total detail in this paper. Many researchers and practitioners have proven over and over again that pellet durability and pelleting efficiency can be substantially improved by the proper steam conditioning of mash. Steam brings to the surface of pellet mash particles the natural oils which are common to most grains and provides lubrication of the pellet die reducing wear on the die and roller assembly and increasing production rates (Behnke, 1990). In some instances, thorough conditioning may be counter productive from the standpoint of pellet durability. If the material slips through the die too easily, dwell time in the die hole is reduced causing the pellet to be less durable, and the starch gelatinisation caused by the heat and friction in the die may be reduced.

Stevens (1987) conducted extensive research into the phenomenon of starch gelatinisation during the feed pelleting process by pelleting corn that was hammermill ground through a 1/8" screen. He used a Perkin-Elmer DSC-23 (Differential Scanning Calorimetre) for gelatinisation analysis. Ground corn before pelleting was used as the control. Samples of the pellets were prepared for analysis in the DSC by grinding them in a Braun coffee grinder, then regrinding in the UDY mill. A 2 mm thick outer portion of pellets was scraped with a razor blade from selected samples and ground in the UDY mill.

The results of the gelatinisation measured in the samples taken immediately after the die are shown.

There was a negative relationship between the conditioned meal temperature and degree of gelatinisation. As the temperature of the conditioned mash was increased, the degree of gelatinisation decreased.

The high degree of gelatinisation that occurred in the outer portion of the pellet at a 23B°C conditioning temperature indicated that heat and mechanical shear next to the surface of the die hole caused a substantial portion of the gelatinisation at all temperatures but, especially, when there were greater temperature differentials between the conditioned meal and the pellet. There is a relationship between that temperature difference and the degree of gelatinisation observed. As the temperature differential decreased, the degree of gelatinisation decreased.

Stevens (ibid) suggested that the conditioning temperature of 80B°C was adequate to gelatinise corn starch; however, the length of time in the pellet mill conditioner at that temperature was probably not adequate for a substantial amount of gelatinisation. It would appear, from that research, that most starch gelatinisation occurred as the feed material passed through the die.

The temperature of conditioning mash has long been a pelleting criterion and an indication of thorough conditioning, that may, or may not, be a totally viable indicator since time at a given mash temperature will affect the conditioning, may affect the degree of gelatinisation, and will certainly affect the pelletability of the mash.

Die Specifications

Behnke (1990) studied the effect of effective die thickness, on length (L), on pellet durability in his experiments reported earlier in this paper. His results indicate, clearly, that durabilities were significantly enhanced with the use of the thicker die; however, production rates were as significantly reduced.

There is no magic. Almost anything that is done to improve pellet quality (durability) will either increase the cost of the ration or reduce the capacity of the pelleting system, or both. Adding to the effective thickness of the die is a perfect example of the sort of trade off that can be expected, and must be recognised, in the search for improved pellet quality.

One of the primary objectives of all commercial feed manufacturers is to economically produce the best pellet quality possible. This is not only important from a customer satisfaction standpoint, but it is becoming apparent that animal performance can be affected by poor quality pellets. Dairy cattle used to consuming pellets, readily reject fines. Even the U.S. broiler integrators are recognising that poor pellet quality can reduce bird performance.

There are numerous factors that affect pellet quality and many are inter-related. It takes a great deal of effort to determine what changes to make and how other aspects of the system or operation might be affected.

Factors not addressed in this paper but that are currently being investigated include: double pelleting, optimum cooling, automation of the pelleting system, gentler handling of pellets and new binders. This paper has not dealt with issues of water stability of pellet aquatic diets, but that topic is gaining great importance around the world.

As can be seen, pelleting is a very complex issue and one that deserves a good deal of thought and investigation.

The Effects of Diet Particle Size on Animal Performance

Cereal grains are the primary energy source in swine and poultry diets. Therefore, not only must producers be concerned about the composition of the grain, but also how it is processed so the animal may fully utilise the nutrients. Since feed represents 65 to 75 percent of overall production costs in a swine or poultry operation, improving the efficiency of feed utilisation will have a tremendous impact on the cost of production. Nearly all feed ingredients will be subjected to some type of particle size reduction.

Particle size reduction increases the surface area of the grain, thus allowing for greater interaction with digestive enzymes. It also improves the ease of handling and the mixing characteristics. However, fine grinding will increase the energy costs of feed processing and may result in feed bridging, dust problems, and increase the incidence of gastric ulcers in swine. Therefore, the increased costs of fine grinding must be offset by improved feed conversion.

Particle Size and Pig Performance: In the past, there has been confusion regarding the optimum particle size in swine diets. This was a result of broad generalisations classifying dietary particle size. In the past, terms like fine, medium, and coarse were used to define particle size. Recently, a more precise classification of determining particle size has been developed based on the mean geometric diametre of particles measured in microns and the geometric mean standard deviation of the particles or their distribution (ASAE 1973). These measurements allow more precise definition of particle size and allow us to make specific recommendations to optimise swine performance.

Complicating the data available on the effects of particle size on pig performance are the interactions between age of the pig, grain type, and particle size. In general, it appears the young pig does a better job of chewing its feed than growing-finishing hogs. The greatest potential for fine grinding to improve feed efficiency will be for finishing pigs. However, fine grinding or rolling will improve feed efficiency regardless of age. A study conducted at Kansas State University demonstrates the effects of particle size on starter pig performance. In the study, 192 pigs (initial weight 13 to 18 pounds) were fed either corn- or sorghum-based diets.

The grains were either processed through a hammermill equipped with an 1/8 inch (539 to 624 microns) or 1/4 inch (722 to 877 microns) screen. Each grain was rolled either fine (822 to 885 microns) or coarse (1,147 to 1,217 microns) by adjusting the roller gap and feeding rate.

As expected, fine grinding or rolling reduced mean particle size of the diet. Although the differences in particle size did not affect average daily gain (ADG), feed efficiency (feed/gain) was improved by either fine grinding or rolling. By pooling the data across grain type and processing method and classifying it based only on particle size, the improved feed efficiency appears to be a result of improved nutrient digestibility.

Recently, Healy et al. (1994) evaluated growth performance of pigs weaned at 21 days of age and fed starter diets in which the grain (corn and hard or soft endosperm milo) was ground to 900, 700, 500, or 300 microns.

These results confirm that reducing grain particle size had little improvement on ADG. Average daily feed intake (ADFI) decreased (linear, P less than .08) as particle size of the diet was reduced. Feed efficiency (F/G) improved then became poorer (quadratic, P less than .01) as particle size decreased suggesting an optimum particle size between 500 and 700 microns.

Pigs fed grain ground to 500 microns had a 6 percent improvement in feed efficiency compared to those pigs fed diets containing grain ground to 900 microns. However, production rate (tons of grain ground per hour) was reduced 43 percent by decreasing particle size from 700 to 500 microns.

Also important is the numerical trend for decreased ADG and ADFI and poorer F/G of pigs fed the diets containing grain ground to 300 microns. The decision on optimum diet particle size needs to include assessment of improvements in feed efficiency versus reductions in milling production. These and other data suggest a dietary particle size of approximately 700 microns to optimise both pig performance and milling efficiency.

The type of grain in the diet also will influence the pig's response to particle size reduction. Studies with high-fibre feed ingredients like barley indicate fine grinding of these types of ingredients may greatly improve their feeding value. A study was conducted with finishing pigs fed diets containing barley ground through a hammermill equipped with either a 1/8, 3/16, or 1/4 inch screen or coarsely rolled. Performance of pigs fed these diets was compared to those fed a diet containing milo ground through a 3/16- inch screen.

Pigs fed the barley diet ground through the 1/8 inch screen had similar average daily gain and feed efficiency compared to pigs fed the milo diet. Daily gain and feed efficiency became poorer as particle size of the barley diets increased. These data indicate grinding of fibrous feed ingredients to approximately 700 microns improves their feeding value and may make them more attractive as substitutes for corn and milo.

Because of its high protein content and propensity to become floury, wheat presents some unique processing problems. If ground too fine, wheat can reduce feed intake. Recommendations for optimum particle size for wheat for use in swine diets should be coarser than corn or milo, between 800 and 900 microns. Roller mills with a differential drive produce a uniform particle size and less fines and may be suited for processing wheat in swine diets.

Particle size data also may be confounded by the kernel size of the grain and the screen size or roller mill settings by which it is processed. For example, corn ground through a 3/16 inch screen will have a finer particle size than either milo or wheat because of its larger kernel size. Corn kernels must be fragmented before they can pass through a 3/16 inch screen opening; however, milo or wheat may fall through the opening intact because of their smaller kernel size.

It is difficult to make a specific recommendation for one screen for each type of grain; however, screen size should be adjusted to produce a mean particle size of 700 microns. In addition, other factors such as hammermill revolutions per minute, tip speed, and the number of hammers also will affect what screen size is necessary to produce 700 micron feed.

One of the disadvantages of fine grinding is the increased incidence of gastric ulcers. Finely ground feed is more fluid when mixed with the digestive secretions of the pig's stomach compared to a more coarsely ground feed. As a result, the acids in the stomach have a greater chance of coming into contact with and irritating the esophageal region of the stomach. The frequency of ulceration increases when particle size drops below 500 microns. Other disadvantages of fine grinding include bridging problems in bulk bins and feeders as well as increased dustiness of the feed.

Influence of Particle Size in Lactation Diets

Many producers believe that finely ground feeds are unpalatable and will decrease feed consumption by the lactating sow resulting in lower litter weaning weight and increased sow weight loss. In order to determine the appropriate particle size for the lactating sow, a study using 100 first parity sows was conducted at Kansas State University (Wondra, 1993). The same dietary formulation was used for all treatments with the only difference being the particle size of the corn in the diet. Corn was ground to 1,200, 900, 600, or 400 microns by either a roller mill (1,200 microns) or hammermill equipped with a 3/8-, 1/8-, or 3/64 inch screen, respectively. Diet particle sizes were tested to be 1,298, 925, 619, and 476 microns.

Litter size, pig survivability, sow weight loss, and sow back fat loss were not influenced by particle size of the diet. However, litter weaning weight gain increased linearly (P less than .05) as particle size decreased from 1,200 to 400 microns. However, the largest response was observed by reducing particle size from 1,200 to 600 microns (9 percent increase in litter weight gain).

Surprisingly, feed intake actually increased (linear, P less than .05) as particle size was reduced suggesting finely ground feed does not decrease palatability of the diet. Digestibility of dry matter, energy, and protein in the diet all improved linearly as particle size was reduced. The reduction in particle size improved nutrient digestibility since small particles have more surface area available for enzyme digestion in the intestinal tract of the pig. The improvement in digestibility is the reason that litter weaning weights were improved as particle size was reduced.

Reduction in particle size is not without a cost. Stomach ulcers and keratinization are always a concern with finely ground diets. Keratinization is an indication of stomach irritation that may lead to ulcers. Stomachs were scored from 0 to 4 (normal to severe ulcer) and from 0 to 3 (none to severe keratinization). Ulceration and keratinization of the stomach was increased as particle size of the diet was reduced.

These data may have greater implication for problems with ulceration considering that the diets were fed for only the 21 day lactation period. Feeding finely ground diets during gestation and lactation may increase the potential for sow death loss because of gastric ulcers. Another study by Wondra et al. (1992) demonstrated production rates decrease and energy utilisation increases as particle size is reduced. Considering these factors, diets for lactating sows should be ground to a similar mean particle size as growing-finishing pigs of approximately 700 microns.

Particle Size for Poultry Diets

In poultry diets, the effects of diet particle size appear to be confounded with complexity of the diet as well as further processing such as pelleting or crumblising. Cabrera (1994) found no effect of diet particle size (1,000 to 400 microns) on growth performance of broiler chicks fed a complex (added tallow, meat and bone meal, and feather meal) diet fed in a crumblised form. In a second trial, feed efficiency was improved 3 percent by reducing particle size from 1,000 to 500 microns in simple diets fed as a meal form but not in crumblised diets. Therefore, the response to reduced particle (600 to 500 microns) size in broiler chicks appears to be greatest when fed simple (grain-soybean meal) diets in a meal form. Feeding a complex diet in a crumblised form did not appear to require particle size below 1,000 microns. Studies with laying hens suggest no advantages in reducing particle size below 800 microns.

Particle Size and On-Farm Feed Processing

A survey of particle size analysis of feed samples indicates that the majority of producers are possibly losing 3 to 8 percent of their feed utilisation costs because of coarsely ground feed. Of the more than 2,500 samples analysed, only 21 percent of them fall within the 700 to 800 micron particle size range recommended for swine feed.

If there are whole kernels in the diet, this is an indicator that the feed isn't ground fine enough. This is a significant finding because particle size reduction has a tremendous economic effect on a swine operation's profitability. If a 100 sow, farrow-to-finish operation has an average diet cost of $130 per ton and reduces particle size from 1,100 to 750 microns, this would result in a savings of approximately $4,750 per year based on improved feed efficiency.

Particle size should become a routine quality control measure and be incorporated into an on-farm quality control program. Some guidelines for establishing a routine particle size monitoring program should include checking ground grain or one complete diet (grower, finisher, gestation or lactation) at least twice a year and up to every 60 to 90 days for large operations. Swine producers who wish to obtain a particle size analysis of their feed can send a 150 gram (3 cups) sample along with $10 to Kansas State Swine Extension, Room 206 Weber Hall, Kansas State University, Manhattan, KS 66506. There also are several commercial laboratories, as well as on-farm particle size kits that can be used for determining particle size. In analysing results of several experiments evaluating the effects of dietary particle size on pig performance, it is clear that the greatest effect of particle size is on feed efficiency. Below is a regression equation based on results from particle size studies. This equation evaluated a range in diet particle sizes from 400 to 1,200 microns fed to pigs from 120 to 240 lb (r = .61; P less than .01).

$$F/G = Analysed\ Particle\ Size - .000415175 + 3.066333$$

This equation can be used as a tool for producers and feed processors to assess the economic impact of particle size reduction. For example, projected feed/ gain for 700 micron feed would be 3.36 based on the equation. If the particle size was increased to 1,100 microns, the projected feed efficiency would be 3.52 or an increase of about 5 percent. While it is important to emphasize that the desired particle size should be 700 microns, an acceptable range of complete feeds is between 600 and 800 microns.

Improving the efficiency of feed utilisation in swine by obtaining a proper particle size will have a tremendous effect on the cost of production. However, improvements in digestibility and feed conversion must be evaluated against potential problems associated with obtaining fine particle sizes. These include increased energy costs, feed handling problems (bridging), gastric ulcers, and dust, as well as reduced production rates. While a specific circumstance may place greater emphasis on one or more of these characteristics than others, a general recommendation is to produce a mean dietary particle size of approximately 700 microns.

Avoiding Drug Carryover During Feed Processing and Delivery

Drug carryover is a form of feed contamination that may result when the substance in question has been transferred (carried) from an acceptable location or feed to an unacceptable location or feed. Carryover of an animal drug can occur during feed processing, handling, or delivery. The Food and Drug Administration's (FDA) Good Manufacturing Practices (GMPs) provide guidance for medicated feed manufacturers to ensure their products meet the identity, strength, and quality standards that they should with respect to their drug content.

These regulations stipulate adequate procedures be established and followed for all equipment used in the production and distribution of medicated feeds to avoid unsafe contamination of medicated and nonmedicated feeds. Carryover of a Category II drug (one which requires a withdrawal period) into a finishing ration may result in a tissue-residue problem in meat animals. Producers experiencing this problem in their market animals may incur significant financial loss. Carryover of either a Category I drug (one which does not require a withdrawal time) or a Category II drug into a batch of feed intended for an off-label species may create serious problems.

For example, carryover of monensin from cattle feed to horse feed may kill the horse. The intent of this bulletin is to provide feed manufacturers, whether a commercial mill operator or an onfarm feed manufacturer, information that will help them avoid drug carryover. This information should help feed processors improve their product quality, reduce the likelihood of feed contamination, and help ensure safe meat, milk, and eggs destined for human consumption.

Risk Assessment: Meat, milk, and egg producers have many options for obtaining a complete feed ration, as depicted. It is the responsibility of the feed processor (whether commercial or on-farm) to ensure that the correct amount of the desired drug is properly incorporated and that no cross-contamination of an unwanted/ unspecified drug is present in that feed. Whether the manufacturing system is simple or complex, it is possible to avoid drug carryover by following the GMPs. The type of drug (Category I or II), number of species, and feed delivery system determine the degree of risk associated with drug carryover.

Feed processors who manufacture products for one species and use only Category I drugs experience the least amount of risk associated with cross-contamination and tissue residue. Since there is no withdrawal time associated with this product, it may be used until the time of slaughter. The individual who uses a Category II drug can avoid carryover and tissue residue problems by using separate delivery systems for these feeds and by sequencing, flushing, and cleaning feed processing equipment. Mills that produce feed for multiple species experience an increased risk of cross contamination by medicated articles (either Category I or II) than mills manufacturing products for one species. When changing feed rations from one species to the next the risk of cross-contamination is minimised through sequencing, flushing, and equipment cleanout of feed processing and delivery equipment.

Causes for Carryover Drug carryover may occur for many reasons as outlined. Significant amounts of the drugs or medicated feed may remain in the production system and contaminate the following batches of feed.

It may occur in one piece of equipment, or it may result from a combination of residues throughout the entire system. Once the source of carryover is known, corrective action can be taken. Often, adjustment of equipment will markedly relieve the problem. Repairs, remodelling, or replacement of worn components may be necessary.

Table 9: Sources of Carry-over

Equipment	*Mode of Carry-Over*
Dust system	-delayed return of dust to production line -excessive pickup of drug and carrier -hang-up (electrostatic or moisture)
Mixer	-residual feed remaining in mixer -buildup of material on ribbons and walls -electrostatic hang-up on walls and top -leaking mixer gate (not fully closed)
Surge bin	-incomplete clean-out -electrostatic or moisture hang-up
Conveyors	-same as surge bin
Elevators	-residual feed remaining in buckets and boot -electrostatic or moisture hang-up
Bins	-bridging -residual feed from incomplete cleanout
Bulk truck	-error in bin chart records -incomplete clean-out -bridging and hang-up

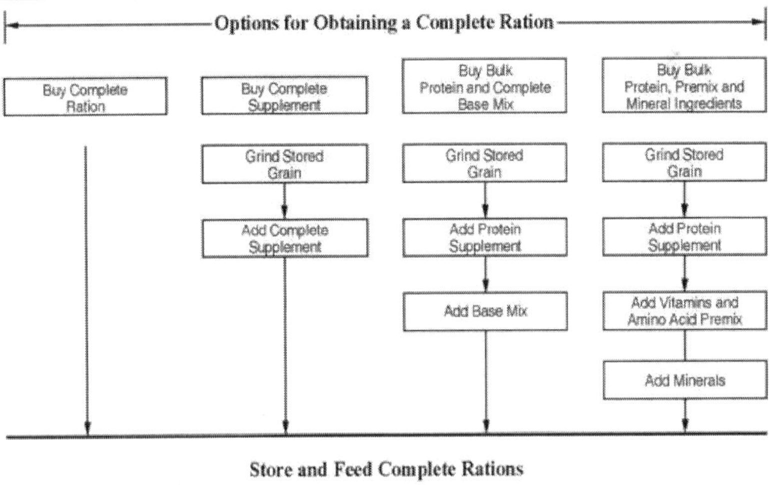

Figure 6: Different Methods Producers can use to Prepare Swine Rations

Avoiding Drug Carryover During Feed Processing and Delivery

Sequencing, Flushing and Equipment Clean-out When working with a drug that requires a withdrawal time before the meat animal goes to market or when manufacturing and delivering feed for several animal species, one must be careful to follow the label and use GMPs to avoid drug cross-contamination. The GMPs state that adequate procedures shall be established and used for all equipment used in the production and distribution of medicated feeds to avoid unsafe contamination of medicated and nonmedicated feed. Three techniques to avoid cross contamination include sequencing, flushing, and equipment clean-out.

The ordering sequence in which feed rations are processed and delivered determines the likelihood of drug carryover and tissue residue. It is an excellent practice to schedule the production of all medicated feeds having the same drug(s) in sequence with the higher levels first and ending with a low level. This sequence should be followed by a nonmedicated feed for the same animal. Individuals manufacturing feed for a single species such as swine, in which a withdrawal drug is fed to young animals, should generally mix feed in the following order: nursery ration containing the withdrawal drug, sow feed, grower, and finishing ration.

Place cull sows in the finishing pen prior to sending them to market if this sequence is followed. When using a sequencing pattern to avoid cross-contamination, it is imperative that feed production records are kept and are detailed enough to denote the last batch/ration. Otherwise, the sequencing pattern could be violated by the next individual preparing feed. In most feed mills, sequencing feed will reduce carryover enough to eliminate the potential for tissue residue. However, sequencing may not reduce carryover to a sufficiently low level if maintenance or design problems exist in the mill as described below:

Table 10: Some Common Corrective Actions for Carryover

Mode of Carryover	*Corrective Actions*
Electrostatic hang-up	-ground wire to affected equipment-purchase nonelectrostatic form of premix-use liquid ingredient to control dust-vibrators to shake hang-up loose
Delayed or extended	-adjust air velocity at collection points-allow more

dust return	time for dust to clear system-use liquid ingredient to reduce dustiness-collect and discard dust following production of medicated feeds (or retain for next run of like medicated feed)-remodel dust system
Mixer residues	-adjust ribbons or paddles-install plastic wipers on ribbons-install air sweep jets for cleaning-remodel discharge for more complete cleanout add drug when mixer is 1/2 to 3/4 full (may affect mixing time required for good mix)
Surge bin, conveyor residues	-adjust for more complete cleanout-remodel bin or discharge
Elevator residues	-adjust bucket clearance in boot (if possible)-install air sweep jets-remodel boot for more complete cleanout
Bin residues	-manual inspection and cleaning when changing kind of feed stored-install vibrator or air sweep jets
Pellet mill and dryer residues	-flush blender and dies-adjust dryer for more complete cleanout
Entire system	-use production scheduling-allow time between kinds of feed for manual cleaning of system- use flush material about 5% of mixer capacity, but not less than 200 lbs. (should be established by actual tests)
Bulk truck	-establish cleanout procedure for truck-require a sample from the first product discharged at point of delivery-analyse delivery samples randomly and let driver know that samples are being analysed

Flushing involves taking a known ingredient, usually ground grain, and moving a quantity through the system to 'flush' out any medicated feed that remains. The amount of flush material depends on the system (about 5-10 percent of mixer capacity) but should not be less than 200 pounds of ground grain. Once the material has passed through the feed processing/ conveying system, it must be stored in a separate bin for use in an identical medicated ration.

Flushing a portable grinder-mixer poses several difficulties, since it would either require transporting several hundred pounds of ground grain (in sacks) to the bulk feeder or storing ground grain in a covered container near the bulk feeder. Flushing procedures for this system include the following: - add 300 pounds (or 5 percent of mixer capacity) of ground grain to the mixer through the charging chute (note: to compensate for the addition of 300 pounds of grain used to flush the mixer, deduct that amount from the feed ration):

- run the mixer for 30 to 60 seconds before discharge,
- discharge the flush material into the bulk feeder containing the feed most recently mixed. A simpler option may involve cleaning the mixer by discharging carryover feed out the bottom port in the vertical mixer. Some portable systems do not contain clean out ports; in this instance flushing may be essential.

Equipment clean-out is often the least used, but potentially most effective, method of avoiding drug carryover during feed processing and delivery. Cleaning the mixer, conveying system, pellet cooler, and sack-off bin or delivery truck between runs to remove residual feed is recommended under high risk situations.

These may include the following: working with a high potency form of a drug (making premixes); sequencing cannot be incorporated into the production schedule; feed processing systems have large carryovers between batches (e.g. portable grindermixers); or when physical properties of drugs are such that sequencing and flushing is not sufficient to prevent carryover. The GMPs stipulate that all equipment shall be designed, constructed, installed, and maintained so as to facilitate inspection and use of clean-out procedures. Scheduled cleaning of mixers is required where liquid ingredients (molasses or fat) are added to the feed ration in the mixer.

Segregation During Handling and Delivery Segregation can occur in ingredients and mixed feeds. A number of sites in the processing, handling, and transit of feed and feed components can produce conditions favourable to the segregation process. Segregation of the drug from the medicated feed may lead to carryover and nonuniform concentration of the medication. Both situations may result in violative tissue residue in the market animal. Mixed feeds are subject to segregation due to differences in particle size, shape, and density. For example, feed ingredients can be evenly dispersed following the mixing process, become slightly segregated as they drop into the surge bin, experience some remixing during transport from the surge bin auger to the elevator leg, become segregated as feed is discharged from the leg and undergoes a free-air fall into the holding bin over the pellet mill, and be partially remixed as material is transferred into the conditioning chamber above the pellet mill. When pellets are formed, the segregation process ceases. The preparation of a meal feed ration may experience segregation and partial remixing throughout the entire handling process until the feed is consumed.

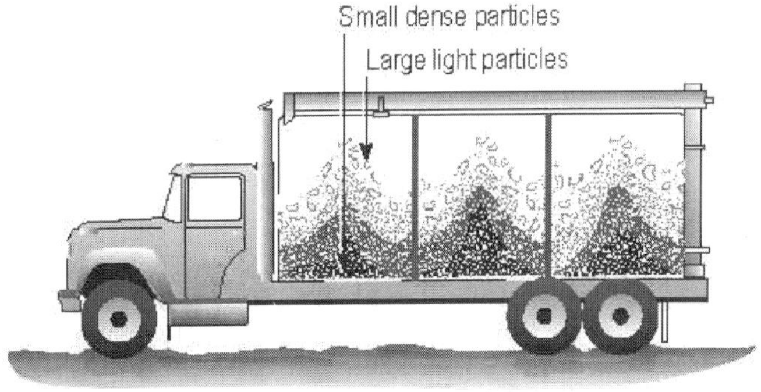

Figure 7: Segregation - Rough Roads

Particles tend to segregate when combined in a complete feed and there is a large size difference between ingredients. Particle shape affects the movement of a material through the air in a freefall situation. Flat particles will tend to fall slower and remain where they fall, whereas particles that are round or cuboidal will fall faster and tend to roll toward the storage wall. Particles with high density will be less affected by free-fall air resistance than those of low density. The less dense particles will tend to be carried toward storage walls by the air currents created in the bin.

Figure 8: Segregation - Free-Air Fall

Management opportunities exist to reduce the amount of segregation that may occur in feed and feed ingredients. Most feed rations contain between 60 and 70 percent ground grain; consequently, the particle size reduction process is critical. Routinely monitoring the grain after grinding will ensure that the desired particle size and uniformity is achieved. Many of the other feed ingredients are delivered in granular form.

Purchasing specifications should be placed on particle size for potential problem ingredients, such as limestone, dicalcium-phosphate, salt, etc. The procedure for evaluating particle size is described in the Kansas State University Extension bulletin MF-2051, Evaluating Particle Size.

Table 11: Sources of Segregation Problems

Sources	***Possible Segregation Problems***
Mixer surge bin	-Free-fall from mixer-Air pressure relief-Mill or equipment vibration-Electrostatic hang-up
Bucket elevator	-Free-fall at elevator discharge
Pneumatic conveying	-Segregation at cyclone collector-Free-fall from collector to bin-Feed angle of repose segregates particles by size
Holding bin, Bagging bin, Bulk bin, Bulk truck, Customer bin	- Free fall through air-Funnelling during discharge accentuates segregation-Vibration of mill or equipment-Electrostatic hang-up
Dust collecting system	-Very fine particles tend to aspirate off-Residue dust is not thoroughly mixed back into feed

Table: Overcoming Segregation Tendencies of Mixed Feed

Property	***Remedy***
Particle size	-Grind ingredients to a uniform particle size range- Use liquids to agglomerate
Particle shape	-Process to more uniform particle shape range-If shape is a desired characteristic, use fat or molasses to agglomerate
Particle density	-Agglomeration is the most common way to overcome density differences-Finer particle size reduces tendencies to segregate because of density differences

Another processing method that will reduce ingredient segregation involves the production of feed pellets. Although a pellet mill is less common for onfarm feed processors due to the capital and operating costs, individuals purchasing complete feed (or processing over 20,000 tons of feed annually) may consider this option. Another commonly used technique to reduce segregation is to add a liquid, usually molasses, fat, or water, to the feed formula. These liquids act to unite small and large particles into agglomerates, which maintain their homogeneity through the subsequent processing and handling. However, liquid addition may create as many problems as it solves. Feed may adhere to equipment and bins creating cleanout and possible feed cross-contamination problems. Proper application, equipment, and location of the spray bar on the nozzle can help avoid these problems.

Drug Carryover and Contamination in Feed Storage and Production

Medicated feeds are important to animal health and growth promotion. Avoiding drug carryover during feed processing and delivery is essential when using medicated articles/feeds. Following the FDA's Good Manufacturing Practices will assist feed processors avoid cross-contamination and help ensure the production of safe meat, milk, and eggs destined for human consumption.

Procedures to avoid cross-contamination between feed batches include assessing the risk and potential causes for drug carryover, preventative maintenance of feed processing equipment, and correct use of sequencing, flushing, and equipment clean-out procedures. Ingredients can segregate during feed handling and delivery. The causes and methods for avoiding these problems are discussed in this bulletin.

A Perspective on Mixing and Mix Uniformity: Part 1 - Introduction, Current Situation, Value of Feed

The basic premise used by all nutritionists when formulating rations is that each aliquot (mouthful) of the diet is balanced with respect to the known nutrient requirement of the target animal. The diet must contain the necessary nutrients to support maintenance, growth, production, and health. Feed additives should be present to provide the appropriate level of protection from disease and other maladies. In all cases, the levels must be controlled so as to be neither deficient nor toxic. The question that must be addressed is how well do current feed manufacturing techniques provide that level of nutrient uniformity assumed by the nutritionist?

The basic objective of any feed mixing operation is to obtain a uniform, random mixture of the solid and liquid ingredients in the formula. The equipment used is, at least in theory, designed to accomplish that objective without nutrient destruction in a minimum amount of time.

A uniform random mixture can only be obtained if there is no favoured direction of movement by individual particles and if there are no selective forces (i.e. centrifugal forces) that come into play. In bulk solids mixing, it is logical that motion must be introduced so that particles are displaced relative to one another.

That is a complex way of saying that if solids are layered one on top of another and no motion takes place, mixing will not occur. However, if the container is rolled, shaken, or vibrated, particle displacement will occur and random uniformity will eventually be obtained.

There are obviously many factors that influence mixing and feed uniformity. They can be divided into ingredient characteristics and machine characteristics. For the time being, only bulk solids will be discussed β€" liquids present special circumstances. The purpose of this paper is to provide a discussion of the current situation and provide guidance as to how feed uniformity can be measured.

Current Situation

Mixing is one of the most important operations in the process of feed manufacturing, yet it is frequently given little or no consideration. The objective in mixing is to create a completely homogeneous blend. In other words, every sample taken should be identical in nutrient (attribute) content to any other sample. Needless to say we seldom achieve that goal, however, we do try to manufacture feed that is as uniform as possible.

Too many times too little emphasis is placed on the mixer we use. When we manufacture a feed, we formulate the diet to provide certain nutrients; in fact, we guarantee that many or all of these nutrients are there in the amounts specified. Many thousands of dollars are spent to gather, process, and store ingredients in semi- or fully-automated proportioning systems to feed exact amounts of ingredients to the scale. Yet, if these varied ingredients are not properly mixed, the quality control system prior to that point will lose a great deal of effectiveness.

Value of Feed Uniformity

Intuitively, the concept of feed uniformity is important and people associated with livestock production realise that if feed ingredients, particularly micro-ingredients such as vitamins, amino acids, trace elements, and drugs, are not properly blended, animal performance will be reduced. Conversely, it is possible to create a toxic situation if some ingredients are not properly incorporated. A recent experience involving urea toxicity comes to mind in which 24 of 25 cattle died because urea segregated from an otherwise safe feed. Most feed additives will fail to provide protection if not properly blended in the feed. Logically, the value of uniformity is greater for the very young animal and animals with a short digestive tract versus older animals that consume large meals less often.

For example, a day-old chick has depleted essentially all nutrient reserves prior to hatching. During his first day on feed, he will consume about ten (10) grams of feed. It is critical that all required nutrients be in that first day ration. As an interesting exercise, one can calculate how many 10 gram aliquots are in a three ton batch of starter feed and then contemplate the probability of each being precisely right. Conversely feed uniformity is not likely to be as critical to a finishing steer that is consuming 12 Kg of feed and has a 24 to 48 hour retention time in his GI tract.

Regardless of the target animal, good manufacturing practices dictate that we strive to produce as uniform of feed as is possible. To that end, equipment should be selected based upon known compatibility. Operational protocols should be set to insure that maximum uniformity is obtained. Personnel should be trained and educated on the concept of uniformity and appropriate testing should be conducted to ensure that uniformity objectives are met.

Equipment Properties

The range of equipment used to mix feed is at least as diverse as the ingredients. There have been many attempts to reduce mixing concepts to a series of engineering equations thus facilitating equipment design from a theoretical approach. The fact is most contemporary mixing equipment, including horizontal ribbon mixers, vertical screw mixers, and drum mixers, have simply evolved from historically successful designs without benefit of mixing theory input. For example, most horizontal mixers have a length approximately three times their diametre and have a rotational speed of 75-100 metres/minute regardless of the diametre. The inside ribbon is usually 2.5 times the thickness of the outside ribbon to balance the directional forces applied due to ribbon diametre.

Given this discussion, it is easy to appreciate the complexity of the mixing operation in a production facility. Yet, this seems to be an area of little concern to most feed manufacturers β€” commercial or private. As regulatory pressures for additive uniformity increases and as the need for providing uniform nutrient density to genetically superior livestock becomes necessary, it will be in the best interest of feed manufacturers to ensure uniformity through testing.

Current and Future Aspects

To focus on the regulatory aspect of uniformity, the following excerpt is taken from the 1990 FDA Regulatory Guidelines (FDA, 1990):

Equipment (225.30) β€” All equipment used in the manufacture of medicated feed shall have the capacity and capability to produce a homogeneous medicated feed of the intended potency. The capability of the mixing equipment should be demonstrated upon installation and periodically as needed to ensure proper adjustments during operation. Written documentation of the adequacy of the equipment should be available for FDA review.

In January of 1990, the Degussa Corporation introduced a program to monitor uniformity of feeds manufactured by customers using their amino acid and other products (Wicker and Poole, 1991). Their results would indicate that only about half of the feeds tested would be of satisfactory uniformity (C.V. < 10%). About 30% had a C.V. of 10-20% and the remaining 20% of the feed samples had a C.V. of > 30%. It is not known precisely at what level of uniformity animal performance will be affected, but one can certainly assume that at a C.V. of greater than 20%, performance would be decreased.

The samples tested in this study were generally from large, centrally controlled feed mills. To gain a perspective on how well on-farm feed manufacturers do, Stark et al. (1991) conducted a study similar to that of Wicker and Poole except using salt as the tracer rather than synthetic amino acids. The results tended to parallel the Degussa report with about 42% of the samples having a C.V. of < 10%, 46% between a C.V. of 10% and 20%, and 12% having a C.V. of > 20%.

It is apparent that, at least in a significant portion of feed produced, nutrient uniformity criteria is not being met. As regulatory authorities move toward required equipment validation, it is imperative that the feed and livestock industries come to agreement as to what levels of nutrient uniformity is needed and how that uniformity is to be measured. There currently exists a standard (ASAE Standards, 1990) for testing solids-mixing equipment for animal feeds; however, the procedure is complicated and a great deal of the data required is meaningless to either regulators or animal performance.

Mixer Testing

The objectives of mixer testing can be many. The most common reason for testing a mixer is to determine the mixing time at which an adequate or satisfactory blend is obtained. The procedures are relatively simple and involve taking samples at specific time intervals. The assay used and statistical treatment is relatively straightforward.

Result Interpretation and Statistical Evaluation of Mixing Tests

The standard deviation and coefficient of variation can be used to measure the results of a mixing test. In very simple terms, they help measure the distribution of values and express the value as one number. In order to interpret the results of a test, the variation of the procedure itself should be known. For instance, the coefficient of variation of the Quantab method is about 10%; therefore, if the result of the mixer analysis is 10% or less, we assume that a Agood mix has been achieved. The same situation applies for other procedures.

One can also use these procedures to isolate points of segregation in the feed mill. If one has a C.V. of 8% at the mixer and a C.V. of 18% after a transfer conveyor, there is a problem between the mixer and that point.

Assay Selection

Numerous assay methods have been used for mixer evaluation. One criteria for selection should be to assay for an ingredient, nutrient, or chemical that comes from a single source. Salt, therefore, is a good selection while protein or nitrogen would be a poor selection.

Sampling is also very important to any mixer evaluation. At least 10 samples should be taken, either from several locations in the mixer or at timed intervals during the mixer discharge. Enough of each sample should be taken each time (approximately 2 lb) to allow for a good test sample. In order to help clarify the requirements for a test procedure to be used for uniformity, the following criteria are offered:

1. The assay principle should be based upon a common ingredient or nutrient that is usually in the formula or can be added without risk.
2. The cost for each assay should be minimal (< $2.00 each).
3. The assay procedure should be simple, fast, accurate, precise, and able to be done on-site.
4. The assay should present no safety hazard to personnel or animals.
5. The assay principle should be supplied from a single source.
6. Sample size required should be reasonable but large enough to reduce or eliminate sampling error.
7. The target mix uniformity (C.V.) should be approximately two (2) times the proven analytical variation for the assay selected but in no case exceed 10%.
8. The statistical procedures required should be easily understood and performed.

As the reader can imagine, there is no "perfect" procedure available.

Mixing Equipment

There are three basic types of mixers used in the feed blending operation: horizontal batch, vertical batch, and continuous. The horizontal batch mixer is by far the most common piece of blending equipment used. The vertical screw mixer is sometimes found in smaller feed mills and is quite frequently used on farms. They are generally slower than a comparably sized horizontal blender.

The continuous blenders vary widely in design and capacity. They are often used to blend liquids into a base mix as well as in the steam conditioning of a pellet mash. Additionally, the continuous mixers are found in plants where continuous rather than batch proportioning is used to arrive at a final homogeneous blend of ingredients.

Horizontal Batch Mixer

This blender is generally equipped with right and left hand ribbon flights which convey material from one end to the other while it is tumbling in the mixer. Other designs include batch paddle mixers in which plows or paddles replace the ribbons. By using paddles, an even higher percentage of liquids may be added at the mixer. The plows are adjustable so that pitch can be set to obtain the desired blending operation.

Horizontal mixers are frequently equipped with multiple discharge openings or with a bottom, which opens along the entire length to facilitate rapid and complete cleanout. They may also be equipped with a surge bin beneath the mixer to reduce emptying and cycle time.

The major advantages of the horizontal mixer over the vertical mixer are the shorter mixing time and the fact that higher percentages of liquid may be added to the blend.

Short-Cycle Mixer

The current generation of mixers are often referred to as Ashort-cycle mixers because they are so effective at mixing that a 1.0 minute mixing cycle or less can be obtained. These mixers generally have twin mixing rotors or twin shafts with paddles. Studies conducted at Kansas State on such a mixer (Forberg) found that a simple ration had a CV of less than 5% after only 15 seconds of mix time. A more complex formula with up to 5% liquid addition will mix in less than one minute.

If a short cycle mixer is to be retrofitted into an existing feed mill, caution should be taken to ensure that the batching system is capable of a cycle time less than the anticipated mixing cycle.

These mixers are more expensive than the more typical ribbon mixers but have attributes that are attractive in many situations.

Problem Ingredients

The physical properties of ingredients can affect mixing. If all of the physical properties are relatively the same, then mixing becomes fairly simple. As the physical characteristics of ingredients begin to vary widely, blending and segregation problems are compounded. Some of these factors are:

1. Particle size
2. Particle shape
3. Density, or specific weight
4. Hygroscopicity
5. Static charge
6. Adhesiveness

Of those listed, particle size, shape, and density are the most important. Large and small particles do not mix well. They will mix better if there are particles of varying size in between. High density particles, such as minerals, tend to segregate and go to the bottom of the bin. An illustration would be the blending of cracked corn and fine mixing salt. The corn may have a particle size of 1200-1500 microns (m) and a density of 1.35 g/cc while the salt might have a particle size in the range of 200 m with a density of 2.6 g/cc. While it is not altogether impossible to obtain and maintain a satisfactory blend of these two ingredients, it is difficult.

Hygroscopicity refers to the ability to attract or take-up water. A very hygroscopic material, such as urea, may absorb enough water from the atmosphere to cause caking. Some ingredients, particularly vitamins and drugs, may possess a large static charge making them tend to stick to bin walls and sides of mixers. As the number of ingredients and variations increases, a greater appreciation of the complexity of the mixing operation is formed.

One of the places we can control ingredient variation is ahead of the mixing operation. Are the grinders, either roller mill or hammermill, producing the desired particle sizes? Is the particle size from the grinders ever measured? We may also need to measure the particle size of ingredients that are normally utilised as received. In some cases they may be contributing to poor mixing.

Discharge and Cleanout of Mixers

There are obvious advantages to the drop-bottom type horizontal mixer in terms of discharge and cleanout. Discharge is nearly instantaneous with the lighter material being discharged within two seconds of the heavier material. There is essentially no carryover from one batch to the next, especially if equipped with air sweep systems.

Liquid Addition

The incorporation of fats, oils, molasses, and other liquids into the vertical mixer is being done today. Many fats and oils are hand-dumped along with the feed ingredients into the dump-in hopper at the base of these mixers. A better way to introduce fats and oils would be through a spray bar installed in the top of the mixer and the liquid metred thereto. The mixes incorporating liquids should be mixed longer than the dry feeds to ensure uniformity and to break up fat or molasses balls that form. To facilitate better mixing and breaking up, a high speed paddle type mixer can be installed in one of the discharge parts and the feed re-elevated to the top of the mixer.

The incorporation of molasses into a vertical mixer is not recommended above 3% if only one mixer is used for all feeds. Even this percentage will require more cleanup maintenance. This requires the installation of access panels in the tub and into the elevating tube to permit cleaning. Molasses builds up on the interior of the tub as well as on the mixing flights and will seriously reduce mixer efficiency. If not kept clean, rejections by the state feed inspection authorities and customer complaints will occur. Teflon lining of the tub and stainless steel screws will reduce maintenance when molasses is being added to the feed mix.

When higher percentages of molasses (3-8%) are going to be mixed continuously in a mixer, then two mixers should be used; one for feed requiring molasses addition and the other for dry feeds.

Liquid addition is best accomplished at the custom mix plant by working on the batch principle. Metring devices can be preset for automatically delivering exact quantities to the mix. The mixers should then operate long enough to properly incorporate or blend the total batch. Attempts to use a continuous, auxiliary mixer for molasses while discharging from the batch mixer can result in headaches. This procedure must be planned very carefully for good results and conditions must be ideal. Getting those conditions ideal is difficult and, for a quantity of molasses (ordered by a customer), the feed man may experience variations in the percentages of molasses applied to the feed as well as a shortage of molasses at the beginning and at the end of the run. If his customers are in the habit of specifying a certain amount, then the feed man should be aware of the continuous system or he may lose customers or give molasses away, thus cutting into his operating profits.

Liquids can be handled equally well in drop-bottom or nondrop-bottom horizontal mixers. About 10% of added liquids are normally considered in the top range of practical use. More than this amount begins to add problems in other areas such as your materials handling equipment. Adding liquids at the batch mixer is a common practice in many feed manufacturing operations. Adding liquids at this point can add to the contamination problem whether they be a drop-bottom or standard discharge type mixer. The addition of liquids (other than molasses) slightly increases horsepower requirements. Regulatory requirements may create a new concept in importance of where and when to add liquids. Where conflicting drug additives follow each other in mixing sequences, contamination is a major problem.

Compliance

Feed manufacturers should not have difficulty meeting requirements established by the regulatory agencies if they adhere to rigid cleaning and operational procedures. When micro-ingredients and premixes are being added, dust collection systems or other air suction systems should be carefully balanced such that a minimum of air is drawn from the mixer. Fortunately, regulations have, in the past, provided a broad range of concentrations in between which the feed manufacturer may operate. However, in light of today improved detection techniques and consumer concern and awareness of potentially harmful drug residue problems, such allowable variations are being reduced. Most feed producers will not be satisfied with producing a feed having as broad a range as was once permitted.

The mixer should, of course, have good mixing efficiency, and allowing for sufficient time to thoroughly blend the feed is an absolute MUST.

When using antibiotics and other micro-ingredients in mixes which are prepared intermittently for poultry, swine, sheep, and cattle, the manufacturer should be sure:

1. That the mixer internals are clean and that it is cleaned out thoroughly between the mixing of the different rations. This means stopping the mixer and removing excess materials in filler tubes and cut-in hoppers.
2. That the mixer is grounded to bleed off static electricity.
3. That regulatory requirements are followed for:
    a. Cleanout of accessory equipment
    b. Labelling
    c. Recheck of ingredient weight and inclusion
    d. Sampling, etc

Operational Comparison of Standard Horizontal Batch Mixers and Drop- Bottom Horizontal Batch Mixers

Operational procedures of the drop-bottom unit is very much like the operation of a normal single or multiple discharge mixer, but it has an added advantage in that it discharges most rapidly; as fast as ten seconds for opening, discharging, and closing of doors. The mixer is completely emptied and clean with very little possibility of contamination. This rapid discharge feature reduces the total mixing cycle time, and, since the feed mill capacity is based largely on mixing capacity, can possibly increase plant capacity through savings in discharge time.

Sequence of Ingredient Addition to Drop-bottom Mixers

We know that there is a decided advantage in following a particular sequence when charging ingredients into the mixer. With the drop-bottom feature you do not have discharge pockets that can possibly catch a few pounds of the first ingredient that is put into the mixer and hold it unmixed. There are no pockets, but a smoothly contoured bottom formed to the contour of the ribbon itself. This further reduces the possibility of segregation and β€œdead spots€□.

Cleaning and Maintenance Requirements

Periodic cleaning of the drop-bottom mixer is no more difficult than the normal periodic cleaning of a standard single or multiple discharge mixer. This can be accomplished through the top or by installing access panels in the side of the catch hopper below the mixer and cleaning it from the bottom with the doors swinging open. Normal cleaning of this unit in dry feed mixing consists of removal of strings that might catch on the ribbon assembly. The only other additional maintenance that would be required would be on the air cylinders themselves and, with proper installation of air supplies with traps and filters, this should be no problem at all.

Noncontamination Feature

The self-cleaning non-contamination feature is a basic point in favour of this particular type of unit. The possibility of contamination must always be considered when using drugs, and we know there are many places in the feed plant where contamination can occur. By eliminating the problem here, one contamination worry is removed.

I might mention, going a bit further, that if you are using bucket elevators to elevate your mixed feed, you can eliminate another source of contamination by equipping your elevator legs with automatic boot cleanout units. These, too, can be integrated into your semi- or fully-automatic batching system to cycle between each batch if desired.

Another way to reduce contamination is to use a gravity flow system wherever possible. With todayβ high energy specialised feeds, with the necessity for feed mill efficiency, and with contamination being the problem it is, we feel that the drop-bottom mixer will prove more and more important in modern feed manufacturing operations.

Nutrient uniformity is the common objective of every feed manufacturer. A great deal of emphasis is placed on uniformity because of regulatory compliance, animal performance, and the satisfaction obtained in producing quality feeds. It is surprising how few mixers are actually tested in any manner, let alone on a routine basis. It is hoped that mixer testing will become more commonplace and will be a part of the quality control procedures of every feed producer.

Further research is needed in the design of mixing equipment and additional animal experiments that will result in a better understanding of the value of uniformity and the cost of non-uniformity in livestock production.

Step Grinding for Improved Efficiency of Grain and Meal Products

Step grinding in the simplest terms, is size reduction accomplished in steps or stages, usually incorporating two grinding machines (hammermills, roller mills, pulverisers, or some combination thereof). The primary objective of step grinding is to reduce the cost to produce a ton of fine ground finished product. Additional benefits may include improved control of the particle size distribution (more uniform grind with less oversize and fewer fines), reduced product heating and subsequent moisture loss, a reduction in the maintenance cost per ton of ground material, finer finished products, and greater flexibility in the grinding circuit.

As noted above, step grinding may be accomplished in circuits utilising two machines, though it is certainly possible to β€œstep grindβ€□ using a single machine, or more than two machines. With a single machine, a step grind circuit will either involve batch processing (grind a batch coarse, readjust the grinding machine finer and process again) or a continuous operation with a screening stage returning oversize materials for reprocessing (circulation grinding). The potential benefits of circulation grinding were explored in the March 1994 edition of Feed Management in an article authored by William L. Ritchie titled Increasing the efficiency of particle reduction. This type of system does offer the same potential of reducing energy and improving particle size control, but does not significantly add to the flexibility of the grinding system.

A second approach, and one that is employed in a number of U.S. feed manufacturing plant is the utilisation of two grinders, one performing a pre-break, and the second grinding the total mixed feed ration. This type of system is commonly referred to as a grinding system, or perhaps just system but differs from the European concept of batching directly to the grinder.

The advantages of this kind of circuit include lower grinding costs, finer finished products, more uniform particle sizing, more uniform finished product mix (lower C.V.), and greater grinding system capacity. The primary disadvantages of this kind of system are the potential for the destruction of some micro ingredients and vitamins, and the higher capital costs to install the system. In most cases, the cost of additional capital equipment is offset in 6 to 12 months in the energy savings of the grinding circuit alone. Additional benefits such as increased (pellet mill) die and roller or (extruder) die life and increased pelleting or extrusion efficiency are bonuses on top of the energy savings.

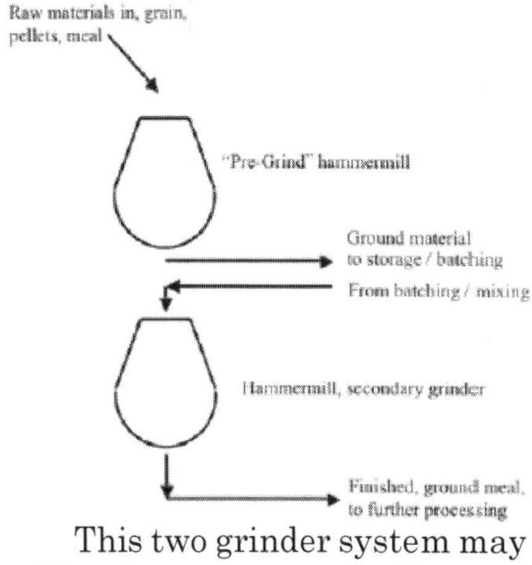

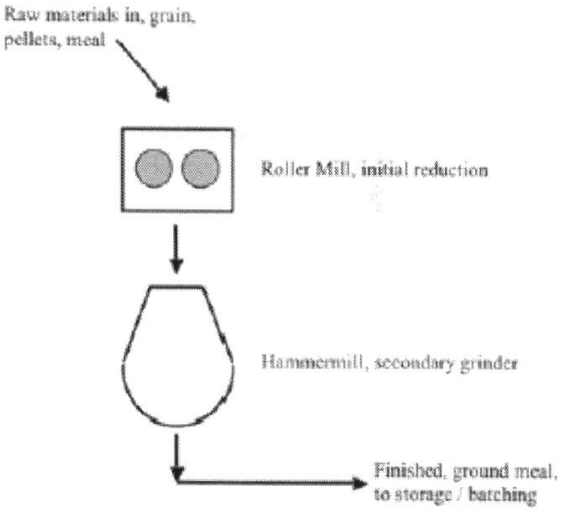

This two grinder system may employ two hammermills, one roller mill and one hammermill, or two roller mills. Additionally, sieving between breaks may be added to further enhance the energy efficiency of the system and reduce operating costs by removing sized materials before the secondary grinder, or by returning oversize materials to the pre-break machine.

Step Grinding, the European Approach

More than ten years ago, a step grinding system approach was being presented by European manufacturers of feed milling equipment as a means of reducing operating costs. An integral part of the European approach was sieving before grinding and sieving between grinding stages. Because the European feed manufacturer uses such a wide range of ingredients received in the form of a meal, there is a potential for a high percentage of the raw materials to already be an acceptable particle size for the feed manufacturing process.

By removing these sized materials, the load on the grinding equipment could be reduced considerably. It appears from research and testing in actual applications that the reduction in energy consumption is roughly equal to one half of the amount of the materials removed. In other words, removing 30% of the materials to be ground (as fines) and by-passing the grinder reduces the energy required to grind by about 15%..

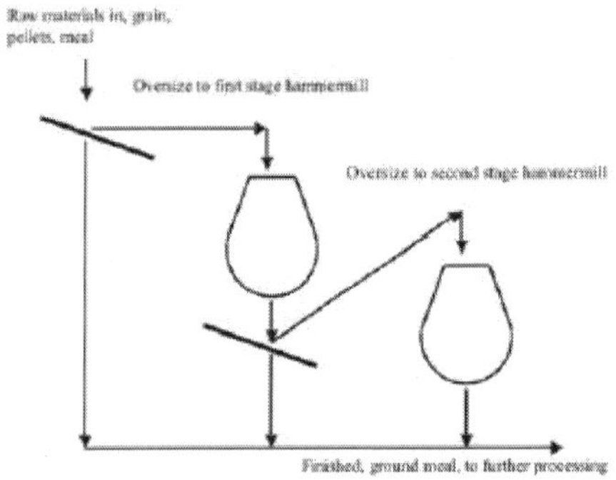

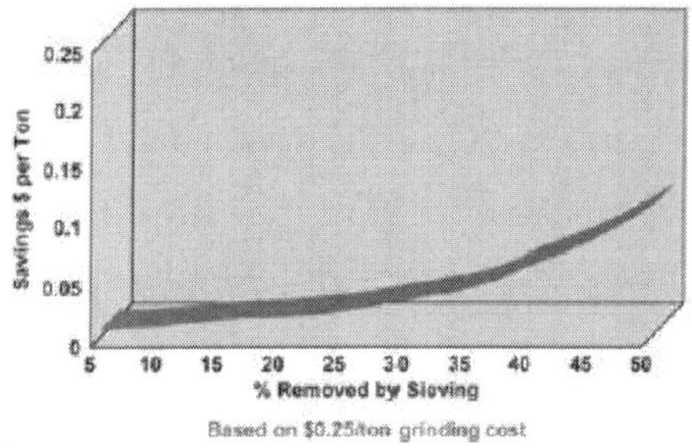

Step Grind for Efficiency in the U.S.

Since corn, wheat, sorghum and barley are the basis for most complete feeds in the U.S., the primary economic benefit of step grinding is an actual reduction in the specific energy required to grind feed materials rather than efficiency gains from sifting. As a rule of thumb for a two shift operation, one horsepower costs approximately \$1.00 per day. An energy reduction of 50 horsepower will save about \$50.00 per day in energy expenditures. Where does this savings come from? Power consumption can be expressed in terms of work accomplished over time.

$$\text{Power} = \frac{\text{Work}}{\text{Time}}$$

The step grind approach yields this energy reduction (less power required) since the materials are reduced more gradually, through a slightly longer period of time. By doing the same amount of work (grinding) over a longer period of time (two or three gradual reductions instead of one instantaneous reduction) the total power requirement is reduced.

Because the roller mill offers such significant energy savings over a hammermill when processing grain in the coarser particle size ranges, the use of the roller mill as a pre-break device can offer substantial savings in a typical grain grinding circuit. By substituting a single pair roller mill in place of the conventional hammermill rotary feeder, the feed rate can be accurately controlled and a significant increase in hammermill capacity can be realised. Because the materials are reduced in size prior to being introduced to the hammermill grinder, larger screens may be employed with no significant increase in finished particle sizing.

This combination of finer feed, increased hammermill efficiency, and the possible use of larger screens not only reduces the energy cost when grinding, but reduces other operating costs (maintenance, parts) as well. Since a rotary feeder is not required the cost of a roller mill for use as a feeder is substantially offset in new installations. In other cases where existing hammermills require a boost in capacity, the application of a roller mill as a feeder can boost hammermill capacity by as much as 40 to 50% with no loss in the fineness of the grind.

Chapter 5

Food Safety and Quality as Affected by Animal Feedstuff

Introduction

In recent years public concern about the safety of foods of animal origin has heightened due to problems arising from bovine spongiform encephalopathy (BSE), dioxin contamination, outbreaks of foodborne bacterial infections, as well as growing concern about veterinary drug residues and microbial resistance to antibiotics.

These problems have drawn attention to feeding practices within the livestock industry and have prompted health professionals and the feed industry to closely scrutinise food quality and safety problems that can arise in foods of animal origin as a result of animal feeding systems. It is important to note, nonetheless, that despite the magnitude of livestock production, the frequency of health problems associated with this sector is very low.

The livestock sector plays an essential role in agricultural and economic development as well as food security. The global livestock output grew at a rate of 2.4% in 1998: increase by over 70% in the next 30 years, and world demand and consumption of livestock products is expected to nearly double in the next 20 years. Most of this increase is expected to take place in developing countries associated with greater population growth and emerging economies, particularly in Asia. Europe has a stable population and, overall, only a modest increase in consumption of livestock products is predicted although notably recent trends in Eastern Europe and the Commonwealth of Independent States (CIS) have shown a modest decline in meat consumption. More detailed analysis of world statistics shows that almost all the growth in demand is in the poultry and pig meat sectors and, even in Europe, there is an increase in white meat consumption and a decline in red meat. This has important implications because it is the poultry, pig and dairy industries that are the principal users of processed animal feeds.

On a global average for 1997, animal products provided about 16% of the calories in the diet. This value increased to 25% in developed countries and dropped to 12% in developing countries. The global value for dietary proteins from animal products over the same period was about 37%, with about 56% in developed countries and 29% in developing countries. Meat and other animal products also provide essential fatty acids, vitamins and minerals. The iron in meat and meat products is easily assimilated by humans and is a key to preventing iron deficiency anaemia, high prevalences of which have been reported in Eastern Europe. The livestock industry therefore can be seen to have great economic and nutritional significance in the world on the whole.

The livestock sector comprises widely differing production systems ranging from subsistence livestock farming to highly intensive systems. The problems related to the quality and safety of foods of animal origin can therefore be highly variable in different regions of the world. This paper discusses linkages between animal feedstuffs and safety of foods of animal origin and highlights steps that have been taken and further steps that are necessary to ensure that animal feed poses no threat to public health. For the purposes of this paper, animal feed includes any substance, whether processed. semi-processed or raw which is used for animal consumption. This includes feed of varied origin such as: pasture, feed grains and compound feeds, crop residues and agro-industrial by-products.

Animal Feeding Practices in Europe

In livestock production, overriding considerations are the availability, quality and efficient use of feed resources, including grassland. In order to meet current demand for meat and animal products, a wide range of systems have been developed to efficiently feed animals. Worldwide, the volume of feed produced annually exceeds 4 billion tonnes.

Animal feeding systems in the wider Europe range from the highly developed systems in the West to less intensive systems especially in rural areas in the East. However, the general model is common and the countries in transition aspire to similar production methods. Poultry and pig production have become almost exclusively intensive in large units. Dairying has followed a similar trend, although grass-based for half or more of the year. Intensive beef production is carried out but this has become increasingly based on grass or maize silage since intensive cereal feeding became less economic during the 1970s. Sheep production is largely extensive and attempts at intensification have proved uneconomic although high stocking rate grazing systems have been sporadically adopted. There are small specialist sectors in goats, milking sheep, deer and buffalo.

Poultry and pigs require protein supplementation and this protein must be of high quality (balanced for essential amino acids) to sustain efficient growth and feed conversion. Selection of improved genotypes has increased the protein and amino acid requirements because of increased protein growth potential and reduced fat. Hence, cereals and other energy feeds must be balanced with additional protein meals. Although these may be of vegetable origin (soyabean, rapeseed, etc.), they require supplementation with amino acids (lysine and methionine) and the best sources (from a nutritional viewpoint) are animal and fish meals. In practice, animal meals are less palatable to pigs and poultry and their use has been limited.

Similarly high producing dairy cows require additional protein supplementation to sustain yields. This is because the cow producing 25 litres or more of milk per day cannot consume sufficient energy and must use body reserves during peak lactation, hence there is not enough energy to provide for microbial protein production in the rumen. Research has shown that this is best supplied with 'by-pass protein', that is protein not degraded in the rumen (also known as undegradable protein). Again, the best sources of 'by-pass protein' are animal and fish by-products. Animal meals were therefore incorporated into dairy concentrates up to the time of the emergence of BSE in the UK.

Gardner reported that the EU was largely self-sufficient in animal feeds, with more than half the annual requirement for feeding its large livestock population coming from grassland and fodder crops. In addition, the EU is a substantial producer of animal feed grains. While being a net exporter of feed grains, the EU is a substantial importer of proteins and non-grain feed ingredients. There is also a continuing trend towards greater intensification of livestock production, particularly in the pig and poultry sectors, leading to an increase in compound feed manufacture and sales. Nutritionally complete feeds are also used in the intensive salmon catfish and marine fish industries of Europe.

The proportion of different ingredients in processed animal feeds in the EU is shown.

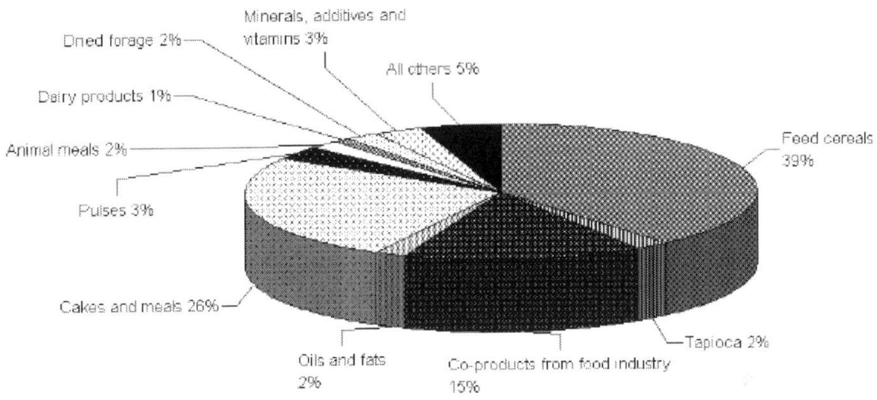

Figure 1: Raw material consumption in industrial compound feeds in the EU-15 in 1996

The growth of industrial livestock has been associated with serious environmental problems. As a result of these concerns, it is predicted that there may be a downsizing of industrial livestock production in some EU countries.

The Netherlands has experienced a trebling of its pig population in the last thirty years and is now reducing numbers (FAOSTAT). This may lead to the growth of intensive systems in the Baltic States and other Eastern European countries with a view to exporting to EU countries to supply the demand.

Potential Hazards and other Issues Associated with Animal Feeds

Mycotoxins

Mycotoxins are secondary metabolites produced by fungi of various generations when they grow on agricultural products before or after harvest or during transportation or storage. Both intrinsic and extrinsic factors influence fungal growth and mycotoxin production on a given substrate. The intrinsic factors include water activity, pH, and redox potential whereas extrinsic factors which influence mycotoxin production are relative humidity, temperature and availability of oxygen.

Many mycotoxins, with different chemical structures and biological activities, have been identified. They may be carcinogenic (e.g. aflatoxin B1, ochratoxin A, fumonisin B1), oestrogenic (zearalenone), neurotoxic (fumonisin B1), nephrotoxic (ochratoxins, citrinin, oosporeine), dermonecrotic (trichothecenes) or immunosuppressive (aflatoxin B1, ochratoxin A, and T-2 toxin). Much of the published information on toxicity concerns studies in experimental animals and these may not reflect their effects in humans and other animals. In addition, the implications for human health of the presence of combinations of mycotoxins are not well understood.

Mycotoxins are regularly found in feed ingredients such as maize, sorghum grain, barley, wheat, rice meal, cottonseed meal, groundnuts and other legumes. Most are relatively stable compounds and are not destroyed by processing of feed and may even be concentrated in screenings. Different animal species metabolise mycotoxins in different ways. For example in pigs, ochratoxin A can undergo entero-hepatic circulation and is eliminated very slowly while it is rapidly excreted by poultry species. The polar mycotoxins, such as fumonisins, tend to be excreted rapidly.

Mycotoxins, or their metabolites, can be detected in meat, visceral organs, milk and eggs. Their concentration in food is usually considerably lower than the levels present in the feed consumed by the animals and unlikely to cause acute intoxications in humans. However residues of carcinogenic mycotoxins, such as aflatoxin B1 and M1, and ochratoxin A, when present in animal products pose a threat to human health, and their levels should be monitored and controlled. The Codex Alimentarius Commission is currently considering maximum limits for Aflatoxin M1 in milk. The extent of mycotoxin accumulation in fish tissues due to consumption of contaminated feed is poorly understood.

Veterinary Drugs

Veterinary drugs may be administered in animal feeds for livestock and aquaculture. If good veterinary practices are employed then Maximum Residue Limits (MRLs) should not be exceeded, however, if GVP is not adhered to, residues in foods of animal origin may exceed MRLs.

Consumer concern about residue levels in foods has fuelled recent trends towards organic and other agricultural production systems that feature reduced reliance on artificial chemical inputs. Foods produced by such systems contain reduced or zero veterinary drug and chemical residue levels. Organic agriculture is one of the fastest growing sectors of agricultural production with an overall persistent growth rate in the EU of around 25% per year for the last ten years.

The need for the containment of anti-microbial resistance due to the use of anti-microbials in livestock, including their addition to feedstuffs, is gaining much attention. Anti-microbials are used for therapeutic, prophylactic or growth purposes, and in the latter case they are added to feed and/or water. Even first-line antimicrobials (e.g. glycopeptides) are being used as feed additives for growth promotion. The assessment and containment of public health risks associated with the use of antimicrobials in livestock is a matter of priority.

Agricultural and other Chemicals

Potential contaminants in feedstuffs include excessive residues of pesticides and fungicides, or other environmental contaminants such as the polychlorinated biphenyls (PCBs), dioxins and heavy metals including mercury, lead, or cadmium.

Dioxins and PCBs are ubiquitously present as contaminants in the environment and dietary intake represents the most common route of human exposure. PCBs and dioxins have similar physical and chemical properties. They are both lipophilic and persistent compounds that accumulate in the food chain, consequently biological samples often contain both dioxin and PCB congeners. Foods of animal origin are the greatest source of human exposure to these contaminants and animal feeds may be an important source of contamination for livestock. Contaminated fats or oils added either intentionally or unintentionally to manufactured feeds can be a source of dioxins and PCBs. These industrial pollutants may be emitted into the air contaminating soil and water and remaining deposited on pastureland. In this case grass-fed animals in highly contaminated areas may give rise to unsafe food products.

Weak associations have been reported between exposure to dioxins and soft tissue carcinomas and lung cancer. Initial symptoms of high PCB exposure are reversible dermal and ocular effects and persistent respiratory problems. Foetal exposure to dioxins and/or PCBs might be associated with cognitive deficits in infants and children. An increase in tumour incidence, as well as neurological, endocrine, hepatoxic and immunotoxic effects were observed in populations accidentally exposed to high levels of PCBs, polychlorinated dibenzofurans and polychlorinated quaterphenyls. Maximum levels of these contaminants allowed in foods of animal origin have been established in some countries, but existing limits are quite variable.

Plant materials growing in areas with high levels of other environmental pollutants such as radionuclides and heavy metals that are used as feed may also lead to unacceptably high levels of contamination in food products of animal origin.

Similarly, fish oils used as animal feed ingredients, may contain high levels of lipid-soluble contaminants if they are produced from fish grown in polluted areas. In Western Europe, effective measures to limit environmental pollution have been put in place. In Central and Eastern Europe some areas of high contamination occur due to industrial activity.

Infectious Agents

Animal feed may be the source of a limited number of infections for farm animals that could lead to human illness on consumption of foods of animal origin. These include Salmonella enterica, Bacillus anthracis, Toxoplasma gondii, Trichinella spiralis and possibly the agent of bovine spongiform encephalopathy. The risk to human health from several other infectious agents which may contaminate feed or forage, appear to be either negligible or non-existent.

Heat treatments of varying severity are commonly used to ensure the microbiological quality of animal feed. Irradiation may be considered a potentially important control measure for certain microbial agents in the feed of food-producing animals.

The Joint FAO/IAEA/WHO Study Group on High Dose Irradiation which convened in Geneva, in September 1997, concluded that food irradiated to any dose appropriate to achieve the intended technological objective is both safe to consume and nutritionally adequate. As their conclusion was partly based on animal feeding studies in a broad cross section of species, fed a variety of diets, it may be assumed that the study group conclusion would also apply to irradiated animal feed. Notably, irradiation is not permitted either by EC Regulation or by Codex Guidelines in organic production systems.

Transmissible Spongiform Encephalopathies (TSEs)

TSEs in ruminants are non-febrile neurological diseases of man and many animal species including ruminants. They have a long incubation period and are ultimately fatal. TSEs are associated with incompletely defined agents, currently termed prions, which are resistant to normal heat treatment of feed and food. Sheep scrapie has been recognised for over 250 years, while Bovine Spongiform Encephalopathy (BSE) was first recognised in the United Kingdom in 1986. For BSE, it has been postulated that the aetiological agent enters the feed primarily through rendered infected tissues (notably the tissue of the central nervous system and the reticuloendothelial system) under conditions of insufficient heat treatment to destroy or inactivate the infectious agent. The reported occurrence of a new variant of the human TSE, Creutzfeldt-Jacob Disease (CJD), has raised the possibility of an association with BSE in cattle with CJD in humans through consumption of meat from BSE infected cattle. At the present time, there is a strong presumption of a link between this new variant and the possible transmission of the infective agent from infected bovine tissue to humans.

Other Infectious Agents

Salmonellae are widely distributed in nature, and animal feed is only one of many sources for farm animals. Animal and plant origin feed ingredients are frequently contaminated with Salmonellae. Processed feed can be contaminated from these raw feed ingredients.

There are over 2000 Salmonella serotypes and these can be divided arbitrarily into three unequally sized groups. These include the species specific serotypes such as S. dublin (cattle); the invasive serotypes which may cause septicaemic disease in several animal species (e.g. S. enteritidis and S. typhimurium); and the non-invasive serotypes which tend not to result in septicaemia. Members of the first group are not recognised as foodborne pathogens. The third group is by far the largest and may be associated with subclinical infections in farm livestock. They can cause disease on occasion and are associated with food poisoning in humans. The principal manifestation of human salmonellosis is a gastroenteritis. Septicaemia occurs in a proportion of patients.

Toxoplasma gondii, the protozoan is found in cats and based on serological surveys also in birds, other domesticated species including sheep, pigs, goats, and horses. The primary source of infection for animals is feedstuffs contaminated with faeces of cats and possibly rodent tissues. A proportion of humans may become infected following the handling or consumption of contaminated raw meat.

Trichinella spiralis is a nematode which parasites the intestinal tract of mammals, particularly pigs. The larvae encyst in the tissues, particularly the muscles which act as a source of infection for humans who consume raw or partially cooked meat. The clinical manifestations include fever, muscle pain, encephalitis, meningitis, myocarditis and rarely death. The cysts can be killed by freezing infected carcasses at -18 degrees C for 20 days. They are also heat sensitive and are killed by traditional rendering temperatures. Effective cooking of raw meat and table scraps before feeding to farm animals would eliminate this hazard.

Bacillus anthracis, which causes anthrax, sporulates on exposure to air and the resulting spores can survive for long periods in the environment and in contaminated animal feed. The spore is widespread in some CIS countries. There have been cases of anthrax in people who have consumed the meat of infected animals.

Genetically Modified Organisms

A new issue concerning animal feed relates to the inclusion of genetically modified organisms (GMOs) in animal feeds. Given that approximately 60% of soy beans planted in the USA are GMOs and this is a principal source for importing into Europe, it is inevitable that they are likely to be present in compound feeds made in the European Region.

This also applies to maize products. While there are no proven harmful effects about the transfer of modified DNA along the food chain, there has been public concern and recent incidents involving protests against this practice. A study by the UK Ministry of Agriculture Fisheries and Food (Report CS0116) was carried out in Leeds University to determine the extent to which DNA in animal feed is degraded under various processing conditions. It addressed concerns over the transfer of foreign genes introduced into GM crops, especially antibiotic resistance marker genes to bacteria, which would render common infectious diseases untreatable. The main conclusions are that DNA is not degraded under most commercial processing conditions, nor in silage, and that further studies on GM material should be undertaken.

Notably, potential feed safety problems related to the use of antibiotic marker genes may be avoidable through the use of marker-free transgenic systems which have now been developed.

Control of Feedborne Hazards

Given the direct links between feed safety and safety of foods of animal origin, it is essential that feed production and manufacture be considered as an integral part of the food production chain. Feed production must therefore be subject, in the same way as food production, to quality assurance including food safety systems based on Hazard Analysis and Critical Control Points (HACCP).

Industry is ultimately responsible for the quality and safety of the food and feed that it produces. National authorities should provide guidance to industry including codes of practice and standards that they must respect. Governments must also establish the necessary controls to ensure that industry consistently meets mandatory quality and safety requirements.

The foregoing outlines the responsibilities of both industry and national governments in ensuring safety of feed and food. It is important to realise, however, that the large volume of international trade in foods of animal origin as well as in feedstuffs adds an important international dimension to the control of animal feedstuffs. Furthermore the World Trade Organisation's (WTO) Agreement on the Application of Sanitary and Phytosanitary Measures (SPS) advocates that national standards related to food safety be harmonised with international standards.

In view of existing shortcomings in the regulation of feed safety, several actions and activities have been undertaken at international level to develop sound standards, guidelines and recommendations in this area. International organisations also have an important role to play in providing information and training which could be used at national level to improve the knowledge and skill of those involved in all areas of the feed industry, including primary producers of feed materials, in order to prevent failures in food/feed safety systems rather than control them.

Work at International Level on the Safety of Animal Feedstuffs

Several international organisations are actively involved in work related to the safety of animal feedstuffs. They generate and disseminate information on various aspects of feeds and their use including potential food safety hazards linked to feed. Furthermore, they provide technical assistance to countries aimed at improving feed production, feeding practices and feed control programmes.

Animal feed safety has become one of the priority areas in the Animal Production and Health Division of FAO which has established a substantial information system on animal feed resources available on the World-Wide-Web, CD-Rom and in printed publications. The system includes detailed scientific and practical information on over 700 feedstuffs, with data sheets, full text articles and reviews. There will be increasing emphasis on feed safety and several reports have been already commissioned on this subject, including one on the use of wastes and by-products.

Interrnational Standards and Guidelines

The Codex Alimentarius Commission (CAC), which was established by the Joint FAO/WHO Food Standards Programme in 1962, is an intergovernmental body comprising 165 members whose responsibility it is to develop and publish international standards, guidelines and codes of practice related to food quality and safety. Codex Alimentarius standards are recognised in the WTO SPS Agreement as the bench marks for food safety.

The issue of animal feed is a complex one requiring multi-disciplinary inputs and collaboration from the fields of human and veterinary medicine, agriculture, academia and from national control agencies. Several organisations are already involved in various aspects of this issue and it is important that there be coordination and discussion among the different parties involved if clear and sound guidance that is widely supported is to be developed in a timely manner. The Intergovernmental Codex Task Force on Good Animal Feeding is an important forum for the required collaboration and consensus.

One of the main tasks facing the Task Force is the completion of the Draft Code of Practice for Good Animal Feeding that was prepared by a panel of experts on the subject during an FAO Expert Consultation that was held in Rome in 1997. The draft code, along with comments submitted by several countries and international organisations will form the basis of the discussions on this topic. All food safety issues identified by the CAC in the terms of reference of the task force can be considered in the elaboration of the new code.

It is considered by some parties that the development of positive and negative lists of feed ingredients may be an important contribution to the control of animal feed internationally.

Antimicrobial Resistance

The matter of antimicrobial resistance is actively being considered by the International Office of Epizootics, which has set up an ad hoc expert group on the topic. Among other issues, this group will consider the development of technical guidelines on the prudent use of antimicrobials and the monitoring of quantities of antimicrobials used in animal husbandry. The ad hoc expert group called upon FAO to take up the role of coordinator with regard to the use of antibiotics as growth promoters.

WHO has held a series of meetings on the question of antimicrobial resistance and based on their findings has recommended the termination of the use of antimicrobials as growth promoters if similar products are also licensed in human medicine. The European Commission Scientific Steering Committee on Antimicrobial Resistance made a similar recommendation in the report of a meeting held in May 1999. The WHO has also prepared draft guidelines for the containment of antimicrobial resistance from antimicrobial use in livestock, which will form the basis of discussions in an expert consultation on the same subject in June 2000. This will have direct implications for animal feeding practices.

The question of microbial resistance is also being discussed within the Codex Committee on Residues of Veterinary Drugs in Foods (CCRVDF) and the Codex Committee on Food Hygiene (CCFH). The former CCRVDF receives scientific advice from the FAO/WHO Joint Expert Committee on Food Additives which currently considers the impact of antimicrobial residues on the gut but does not consider the transfer of antimicrobial resistance arising from the use of antimicrobials and their release into the environment, as this is beyond their terms of reference. The role of CCRVDF in broader discussions of the question of antimicrobial resistance is uncertain. CCFH considers antimicrobial resistant bacteria in relation to food hygiene. This committee is already involved in risk assessment associated with microbiological contamination of food, and is therefore well-placed to take a risk analysis approach to the question of antimicrobial resistance.

Environmental and Industrial Contaminants

The recent problems originating from the contamination of animal feed in Belgium highlighted the existing disparity among regulations in different countries related to the presence of dioxins and PCBs in foods including their maximum allowed limits. The Codex Committee on Food Additives and Contaminants (CCFAC) has requested that the FAO/WHO Joint Expert Committee on Food Additives (JECFA) consider these groups of compounds and provide scientific advice to guide the Committee in its task of establishing guideline or maximum levels in food and feed.

WHO held an expert consultation on assessment of the health risk of dioxins in Geneva in May 1998. The consultation evaluated the tolerable daily dose of dioxins to which a human can be exposed without appreciable health risk. In light of epidemiological data concerning the effects of dioxins at low levels of exposure and based on animal studies, the TDI was reduced from 10pg/kg body weight to a range of 1-4pg/kg body weight.

CCFAC has also initiated work on a proposed draft Code of Practice for Source Directed Measures to Reduce the Contamination of Food with Chemicals. At the 32nd Session of this Committee it was decided to forward this proposed draft to the Executive Committee for preliminary adoption. This work is important in terms of reducing contamination of animal feed which is produced in areas highly affected by environmental pollutants. CCFAC is also currently working to establish maximum limits for the presence of lead and cadmium in food and feed materials.

Programmes to monitor and control levels of environmental and other chemical contaminants in food and feed must be supported by adequate analytical expertise and equipment. The dioxin crisis revealed that until recently, few laboratories were equipped to carry out testing for this group of contaminants. FAO is frequently involved in providing technical assistance to countries to improve their capability in food contaminant analysis. Ongoing FAO Technical Cooperation project in Lithuania and the Slovak Republic involve such aspects of food/feed control.

Mycotoxin Contamination

During its 32nd Session held in Beijing in March 2000, the Codex Committee on Food Additives and Contaminants decided to create a single general Code of Practice for the Prevention of Mycotoxin Contamination in Cereals, including specific Annexes related to the prevention of Ochratoxin A, Fumonisins and Zearalenone contamination in cereals, for consideration at its next meeting. Given that approximately 26% of the world cereal production is used directly for animal feeding (FAOSTAT) this work has direct implications for the control of mycotoxin contamination in feed.

The capability and reliability of food control analytical services are important considerations in the implementation of national programmes to ensure feed safety. FAO, in collaboration with the International Agency for Atomic Energy (IAEA), has been widely involved in improving analytical quality assurance in laboratories carrying out mycotoxin analysis and has implemented several training courses on this topic. Another such training course is planned for the countries of Eastern Europe in October 2000. These training programmes help prepare laboratories to obtain official accreditation.

Transmissable Spongiform Encephalopathies

Research and investigation continues in the area of TSEs due to the important implications for the safety of foods of animal origin. In December 1999 WHO, in collaboration with the International Offices of Epizootics (OIE), held a consultation on Public Health and Animal Transmissable Spongiform Encephalopathies - Epidemiology, Risk and Research Requirements. The recommendations of this consultation included the need for continued priority to be placed on surveillance of Creutzfeldt-Jacob Disease (CJD) and evaluation of animal TSE zoonotic potential; and, continued emphasis on programmes for the control and eventual eradication of TSEs in livestock including the development of improved diagnostic testing methods.

Genetically Modified Organisms

In relation to the use of GMOs in animal feedstuff the main concerns relate to: the possibility of DNA of inserted or modified genes or their products causing adverse health effects in animals; and the possibility of DNA fragments or proteins being transferred to or accumulated in foods of animal origin (milk, eggs and meat) leading to adverse health effects in humans consuming such foods. Research carried out so far has not demonstrated any hazard posed by GMOs, but further work is required.

One of the several important issues facing the Codex Task Force on Good Animal Feeding relates to the use of GMOs in animal feed. The work of the task force for the Safety of Novel Foods and Feeds which has been established by the Organisation for Economic Cooperation and Development (OECD) is also highly relevant to this issue. The question is a complex one and further research will be necessary to resolve some food/feed safety issues. What is essential in the short term, however, it to achieve consensus in the international community concerning the approach to the evaluation and regulation of GMOs in food and feed. This process is already well underway.

There is general agreement that the concept of substantial equivalence elaborated by OECD is the most practical approach to address the safety evaluation of food or feed components produced by modern biotechnology.

The safety evaluation itself should follow risk analysis principles as elaborated by relevant international organisations, including a precautionary approach in cases where there is insufficient knowledge or understanding to support an exhaustive risk assessment. This matter is being considered by the Codex Ad Hoc Inter-Governmental Task Force on Foods Derived From Biotechnology, which held its first session in Japan, 14-17 March 2000. A Joint FAO/WHO Expert Consultation on Foods Derived from Biotechnology was held in Geneva, May/June 2000, to elaborate further on the concept of substantial equivalence and to consider other pertinent issues.

There is also agreement on the consumer's right to be informed of how their food is produced. The Cartagena Protocol requires that GMOs moving in international trade be identified as such. Furthermore, the issue of labelling of foods produced by biotechnology is currently being discussed within the Codex Committee on Food Labelling.

Code of Practice for Good Animal Feed Production, Handling and Storage

The FAO Expert Consultation on Animal Feeding and Food Safety held in 1997 noted that quality assurance (QA) begins with the concept of what the feed product is to be, in terms of the species being fed and the results being sought. Ingredient specifications are important to quality assurance in defining the quality of the feedstuffs to be accepted by the processor when the raw materials are received for processing. The formulation of the finished feed, including any added medications, should meet the regulatory requirements of the government as well as satisfy the animal production objectives of the customer. Other QA factors involve the manufacture and distribution of the feed. Key elements in effective quality assurance at the feed production facility should include proper sampling, laboratory testing and microscopy, in-plant quality control, control of drug carry-over, plant sanitation and integrated pest management, plant cleanliness, the receiving area, and storage. Quality assurance procedures must be clearly documented and records maintained.

Clear guidance from governments to industry regarding required features of feed quality assurance programmes constitutes an important preventative approach to feed safety. Such guidance can be provided through codes of practice for the animal feed and livestock sectors covering all stages of feed production, handling and storage to its ultimate use. The code could cover such issues as: production of feed materials; sourcing of feed materials; use of additives and veterinary medicinal substances; structures, equipment and materials used in feed manufacture, packaging, handling or storage; product and personnel flows through the facilities; processing operations; labelling requirements; transport and storage of finished product; traceability mechanisms; and, appropriate record-keeping.

Applying HACCP ensures that all potential safety hazards are thoroughly analysed and assessed, that critical limits are established for all points along the chain that must be controlled to avoid occurrence of safety hazards, that effective systems for monitoring the critical control points are in place, and that plans for corrective action are established in the event of problems within the production chain. Processors and handlers of animal feed must further ensure that adequate documentation is maintained to demonstrate their adherence to HACCP principles.

The work currently being undertaken at international level to establish a Code of Practice for good animal feeding will be instrumental in ensuring a uniform approach to assuring the quality of feed.

National Programmes to Prevent Foodborne Hazards Related to Animal Feed

A basic component of the national controls necessary to guarantee the quality and safety of feeds and foods includes legislation and regulations. The legislation defines responsibilities and designates authority with respect to the wide spectrum of activities involved, establishes basic procedures to be followed in enforcing the legislation, and provides standards, guidelines and other recommendations to be respected by the industry in the production of animal feed.

Legislation and regulations concerning the production, manufacture, handling, storage and use of animal feed should be coherent and complementary parts of national food legislation. The development and implementation of food law is a political process, which is dynamic and evolutionary and often reflects changing public and political concerns. Unless governments undertake periodic review of food/feed legislation to ensure its coherency, it may become a patch work of additions and revisions leading to overlapping, redundant and conflicting jurisdictions for government agencies and gaps that could give rise to public health problems. Recent food safety problems involving unsafe feed have demonstrated certain weaknesses and gaps in existing legislation in force in many countries.

Some countries may wish to update policies in line with today's food and feed safety issues. For example given the current possibilities of modern biotechnology, it may be necessary that regulations and procedures for the evaluation, authorisation and labelling of foods produced from biotechnology, be established. Measures instituted should not be more restrictive than necessary to meet legitimate food/ feed quality and safety objectives.

The WTO SPS Agreement calls upon members to harmonise food safety measures internationally. Recent food safety crises linked to feed contamination have heightened international awareness of existing disparity among national legal limits for maximum levels of contaminants that could be associated with animal feedstuffs. This raises questions about the appropriate level of consumer protection and also constitutes a barrier to trade. Current work by relevant international organisations to determine appropriate guidelines on the levels of contaminants is necessary to facilitate eventual harmonisation of regulations without compromising public safety.

Enforcement of legislation and regulations depends on effective administration of feed control programmes, the existence of an inspection service whose staff is well trained in feed production and manufacture and related safety issues, analytical services adequately equipped with sufficient capacity and the required technical expertise to carry out the volume and types of analyses necessary to support monitoring and surveillance programmes as well as routine regulatory testing.

There is agreement throughout the international community that food safety measures should be scientifically justified. This highlights the need for research on feed-related safety hazards in foods of animal origin. The establishment of research networks could be a useful strategy to promote sharing of scientific information the improving the chances of building more rapidly on scientific developments to arrive at reliable answers on questions regarding feed safety and quality.

Where potential problems associated with feed are not effectively covered by existing rapid alert systems for food safety, governments should take the necessary steps to ensure the implementation of swift action to contain and eliminate any feed safety problems with the potential to cause public health risk. In the case of feed in international trade, Codex provides guidelines for the Exchange of Information in Food Control Emergency Situations (CAC/GL 19-1995).

In seeking to address the weaknesses in feed control that have become apparent in recent years, governments must pay considerable attention to the inter-relations that exist between different agencies and the need to establish cooperative and coordinating mechanisms to ensure the best possible implementation of all programmes while giving privilege to public health concerns when these arise. The composition and safety of feed have important food safety implications, but are also critical considerations in animal health and husbandry. It is fitting therefore that governments establish suitable mechanisms for food safety agencies, agriculture departments and veterinary services to work together in determining the best policies and actions with regard to animal feed. It must be emphasised that an important aspect of control lies in training. Poor practices within the animal feed industry are often linked to lack of knowledge of associated hazards and how these should be managed. The development of relevant training and extension programmes for people involved in the feed industry or in animal feeding at farm level is an important aspect of control of animal feed.

The control of BSE offers a clear illustration of the need for close collaboration among food control and other concerned agencies. Measures that have been established in Europe and elsewhere centre around surveillance and notification of BSE in live animals, establishment of processing conditions for mammalian tissues to be used in animal feed and prevention of the use of mammalian tissues in feedstuffs for ruminants. The first of these activities would fall under the responsibility of veterinary services. The complexity of the flows of raw material, animal meal and compounds containing animal meal highlights the need for close cooperation between various agencies in order to achieve public health objectives.

Certain chemical substances and biological agents incorporated into feed at any stage of production up to the point of feeding, either intentionally or unintentionally, can result in hazards in food of animal origin. National food safety programmes should therefore include the control foodborne hazards that originate in feedstuffs.

Effective control of animal feedstuffs requires multidisciplinary input. There is a need for collaboration between all parties involved in feed and animal production, especially those in a position to provide veterinary clinical and epidemiological information, to establish the linkage between any identified or potential hazard and the level of risk. Such information is essential for the development and maintenance of appropriate risk management options and safe feeding practices.

The disciplines that apply to international trade in both food and feed, as well as in feed ingredients were agreed upon during the Uruguay Round of Multilateral Trade Negotiations and set out in the SPS Agreement. Considerable work has been undertaken by the international community to support a uniform approach towards ensuring feed safety that is consistent with SPS principles.

Recommendations

Based on the above findings and conclusions, Member Nations could consider adopting the following measures:

- review existing food safety and quality legislation so as to ensure that it provides an adequate basis for the control of feed-related hazards with the potential to cause public health risks;
- participate to the fullest extent possible in the work being undertaken by international organisations involved in developing standards, guidelines and recommendations relating to feed-borne hazards;

- support research into the public health implications of animal feeds to support risk analysis of feedborne hazards; this would facilitate the setting of MRLs and the determination of effective control measures for protecting public health.
- develop suitable analytical and diagnostic methods for rapid screening and confirmation of feedborne hazards in national surveillance and monitoring programmes as well as in routine regulatory testing;
- establish mechanisms to ensure multidisciplinary scientific input, involving human medicine, veterinary medicine, animal science, crop science, food/feed technology, environmental science and toxicology, into policy and programme decisions relating to the control of feedborne hazards;
- encourage dialogue among producers of feed or feed ingredients, livestock and aquaculture industries and government as an essential part of the process of elaborating codes of practice for the feed industry; this will ensure that eventual regulations and guidelines are practicable and widely supported;
- establish, as needed, coordination and cooperation among several government agencies and departments to ensure successful and efficient implementation of feed control programmes;
- establish policies and programmes for the evaluation, authorisation and labelling of feed containing GMOs; these programmes should be no more restrictive than required to meet legitimate objectives.

International organisations should continue to develop and make available information related to animal feed safety to their member countries thus supporting national feed control programmes.

Animal Feeding and Food Safety

Potential Hazards Associated with Feed:

Mycotoxins: Mycotoxins are secondary metabolites produced by fungi of various genera when they grow on agricultural products before or after harvest or during transportation or storage. Some fungi such as *Fusarium* spp. typically infest grains before harvest, others such as *Penicillium* spp. can invade grain after harvest, while *Aspergillus* spp. can grow on grains both before and after harvest. It must be emphasised that the presence of the fungi does not necessarily imply that mycotoxins can be found. Conversely, the absence of fungi does not necessarily mean the absence of mycotoxins.

Both intrinsic and extrinsic factors influence fungal growth and mycotoxin production on a given substrate. The intrinsic factors include water activity, pH, and redox potential whereas extrinsic factors which influence mycotoxin production are relative humidity, temperature and availability of oxygen.

Many mycotoxins, with different chemical structures and biological activities, have been identified. Mycotoxins may be carcinogenic (e.g. aflatoxin B_1, ochratoxin A, fumonisin B_1), oestrogenic (zearalenone and I and J zearalenols), neurotoxic (fumonisin B_1), nephrotoxic (ochratoxins, citrinin, oosporeine), dermonecrotic (trichothecenes) or immunosuppressive (aflatoxin B_1, ochratoxin A, and T-2 toxin). Much of the published information on toxicity concerns studies in experimental animals and these may not reflect their effects in humans and other animals. In addition, the significance of the presence of most mycotoxins in foods of animal origin is not completely understood.

Mycotoxins are regularly found in feed ingredients such as maize, sorghum grain, barley, wheat, rice meal, cottonseed meal, groundnuts and other legumes. Most are relatively stable compounds and are not destroyed by processing of feed and may even be concentrated in screenings.

Different animal species metabolise mycotoxins in different ways. For example in pigs, ochratoxin A can undergo entero-hepatic circulation and is eliminated very slowly while it is rapidly excreted by poultry species. The polar mycotoxins, such as fumonisins, tend to be excreted rapidly.

Mycotoxins, or their metabolites, can be detected in meat, visceral organs, milk and eggs. Their concentration in food is usually considerably lower than the levels present in the feed consumed by the animals and unlikely to cause acute intoxications in humans. However residues of carcinogenic mycotoxins, such as aflatoxin B_1 and M_1, and ochratoxin A, when present in animal products pose a threat to human health, and their levels should be monitored and controlled.

In most instances the principal source of mycotoxins for humans is contaminated cereals and legumes rather than animal products. This means that the exposure to mycotoxins may be greater in developing countries in which cereal grains and legumes form the staple diet and the intake of animal products, including meat, is low.

There is little information available regarding the occurrence of mycotoxin residues in animal products intended for human consumption. Examples of maximum levels in force in various countries include 0.05-1 ppb for aflatoxin M_1, 5 ppb for aflatoxin B_1, 25 ppb and 50 ppb ochratoxin A in porcine kidneys and cereals respectively and, depending on the country, 30-1,000 ppb for zearalenone in corn and foods. The levels of mycotoxins detected are usually below the maximum levels accepted in most countries.

Table 1: Examples of food of animal origin which may be naturally contaminated with mycotoxins

Mycotoxin	*Potential effects on humans*	*Occurrence*	*Maximum level reported*
Aflatoxin B_1	Hepatic cancer	Eggs	0.4 ppb
		Pig liver	0.5 ppb
		Pig muscle	1.04 ppb
		Pig kidney	1.02 ppb
Aflatoxin M_1	1	Cow's milk	0.33 ppb
Ochratoxin A	Renal damage	Pig liver	98 ppb
		Kidney	89 ppb
		Sausages	3.4 ppb
Zeralenone	Oestrogenic	Pig liver	10 ppb
		Pig muscle	10 ppb

There is insufficient evidence to describe aflatoxin M1 as a human carcinogen although it is a potent carcinogen in rodents.

Infectious Agents

Transmissible Spongiform Encephalopathies in Ruminants

The transmissible spongiform encephalopathies (TSEs) are non-febrile neurological diseases. They have a long incubation period and are ultimately fatal. TSEs are associated with incompletely defined agents currently termed prions which are resistant to normal heat treatments of feed and food. The TSEs recognised in food producing animals are BSE and scrapie. Sheep scrapie has been recognised for over 250 years. BSE was first recognised in the UK during 1986. The BSE infectious agent enters feed primarily through infected tissues (notably the central nervous system and the reticuloendothelial system) rendered under conditions of insufficient heat treatment to reduce the concentration of the infectious agent to an ineffective dose.

In the case of sheep scrapie, infection is naturally maintained by transmission between sheep. It is likely that humans have been exposed to the scrapie agent by eating brain and other tissues from infected sheep although there is no evidence that the occurrence of either CJD or nv-CJD has been associated with scrapie. With respect to BSE, humans can potentially be exposed through consumption of the infected tissues. The occurrence in humans of nv-CJD has raised the possibility of an association with the BSE agent. At present, with the limited number of diagnosed cases, there is no proven link between nv-CJD and the possible transmission of the infective agent from bovine tissue to humans.

Salmonella Enterica

There are over 2,000 salmonella serotypes and these can be divided arbitrarily into three unequally sized groups. These include:

1) the species specific serotypes such as *S. dublin* (cattle) and *S. gallinarum* and *S. pullorum* (poultry);
2) the invasive serotypes which may cause septicaemic disease in several animal species (e.g. *S. enteritidis* and *S. typhimurium);* and
3) the non-invasive serotypes which tend not to result in septicaemia. Members of the first group are not recognised as feedborne pathogens.

The third group is by far the largest and may be associated with subclinical infections in farm livestock. Occasionally they can cause disease and are associated with food poisoning in humans. The principal manifestation of human salmonellosis is gastroenteritis. Septicaemia occurs in a proportion of patients. The case mortality rate is low with the young, old or immuno-compromised being most susceptible.

Salmonellae arc widely distributed in nature, and feed is only one of many sources for farm animals. Feed ingredients, of both animal and plant origin, are frequently contaminated with salmonellae although the most common serotypes isolated are rarely the most prevalent in animals including man. The two most important serotypes associated with human disease, *S. enteritidis* and *S. typhimurium,* are rarely isolated from feed. Feed can be contaminated by contact with raw ingredients after processing.

Toxoplasma Gondii

The protozoan *Toxoplasma gondii* is found in cats, and based on serological surveys, in birds, and in domesticated species including sheep, pigs, goats, and horses. The primary source of infection for animals is feeds contaminated with cat faeces.

Cats are an important source of infection for humans, with the handling or consumption of raw meat also being implicated. Pregnant women who become infected may abort or give birth prematurely, and infants often develop central nervous system disorders and ocular disease. Immuno-compromised patients are at particular risk.

Trichinella Spiralis

Trichinella spiralis is a nematode which parasitises the intestinal tract of mammals, particularly pigs. The larvae encyst in the tissues, particularly the muscles, which act as a source of infection for humans who consume raw or undercooked meat. The clinical manifestations include fever, muscle pain, encephalitis, meningitis, myocarditis and very occasionally, death.

The cysts can be killed by freezing infected carcasses at -18°C for 20 days. They are also heat sensitive and are killed by traditional rendering temperatures. Effective cooking of raw meat and table scraps before feeding to farm animals will eliminate this hazard.

Veterinary Drugs and Agricultural and other Chemicals

Veterinary Drugs: Veterinary drugs may be administered in animal feeds. If the concentration used results in foods of animal origin with residues exceeding the established Maximum Residue Limits (MRLs) such as those established by Codex, there may be a potential risk to human health. Codex MRLs should not be exceeded if concentrations used are correct, withholding times are observed and Good Agricultural Practices (GAP) and Good Veterinary Practices (GVP) are applied.

Agricultural and Other Chemicals

The potential hazards may include excessive residue levels of herbicides, pesticides, and fungicides and industrial/environmental or other extraneous contaminants such as the polychlorinated biphenyls (PCBs) and heavy metals including mercury, lead, or cadmium. Cereals and treated seeds are the most likely source of these contaminants. The most significant hazards to human health are those chemicals that accumulate in animal tissues or are excreted in milk or become incorporated in eggs at levels in excess of established limits such as the Codex MRLs for pesticides or maximum levels for contaminants in a food or feed.

Assessment of the Risk

The risk analysis approach has been adopted within the Codex system and used as the fundamental method underlying the development of food safety standards. The Consultation therefore saw one of its tasks as to provide an assessment of the risk of foodborne hazards that enter the food chain via feeds.

Risk analysis is composed of three separate but integrated elements, namely risk assessment, risk management and risk communication. Risk assessment has been defined by Codex as being the scientific evaluation of known or potential adverse effects resulting from human exposure to foodborne hazards. The risk assessment process consists of the following steps: (i) hazard identification, (ii) hazard characterisation, (iii) exposure assessment, and, (iv) risk characterisation. The definition includes quantitative risk assessment, which emphasises reliance on numerical expressions of risk, as well as an indication of the attendant uncertainties. Hazard identification is defined as the identification of known or potential health effects associated with a particular agent. Exposure assessment is the qualitative and/or quantitative evaluation of the degree of intake likely to occur. Risk characterisation is the integration of hazard identification, hazard characterisation and exposure assessment into an estimation of the adverse effects likely to occur in a given population, including attendant uncertainties. For chemical agents, a dose/response assessment should be performed. For biological or physical agents, a dose-response assessment for hazards should be performed if the data are available or obtainable.

In general terms, the Consultation recognised that there are risks arising from foodborne hazards that enter the food chain via feeds. However, on balance, the judgement was that the risk of these hazards was low in comparison to foodborne hazards that originate from other sources. The risks from *Salmonellae* for example, may be considerably greater during processing of carcasses and subsequent animal product processing. The exposure to mycotoxins is far greater from eating contaminated cereal grains than from eating foods derived from animals fed contaminated grains.

Sources of Feedborne Hazards

Feeds can be of animal, plant, microbial or mineral origin. The following tables set out potential hazards and the stage at which they may enter the food chain.

Table: Feed of plant origin

	Pre-harvest	*Post-harvest*	*Post-processing*
Grains and their by-products	A, B, D	A	A
Oilseeds and their by-products	A, B, D	A, B, D	A, B, C, D
Molasses	B, D	-	A (possible)
Roots and tubers	B, D	A	A
Forages	A, B, C, D	A, C	A, B
Roughages	-	A, B, C	-
Fats and oils	-	-	B

Key: A = mycotoxins, B = agricultural chemicals, C = microbial pathogens, D = metals

Table: Feed of animal origin

	Raw Post-processing materials	
Mammalian protein meals	B, C, D, E, F	C
Poultry meals	B, C, D, E	C
Aquatic animal meals	B, C, D	A, C
Fats and oils	B	-

Key: A = mycotoxins, B = agricultural chemicals, C = microbial pathogens, D = metals, E = drug residues, F = TSEs

Table: Miscellaneous feed ingredients

	Raw material	*Post-processing*
Mineral Origin		
Phosphate, calcium and sodium sources	D	-
Trace element premixes	D	-
Non-nutritive adsorbents	D	-
Microbial Origin		
Yeast and single cell protein	C, D	D

Miscellaneous Origin		
Food waste	A, C	A, C
Animal manure	B, C, D	C

Key: A = mycotoxins, B = agricultural chemicals, C = microbial pathogens, D = metals.

Control of Feedborne Hazards

Feed and feed ingredients should be obtained and preserved in a stable condition so as to prevent hazardous effects due to contamination or deterioration. When received, feeds should be in good condition and meet generally accepted quality standards. Preservation can be facilitated by low temperature storage, ensiling, dehydration or the addition of appropriate chemicals (e.g. propionic acid). Furthermore, pasteurisation reduces the numbers of most pathogens. Maintaining low water activity (i.e. A_w <0.65) will minimise bacterial and fungal growth.

Good Manufacturing Practices (GMPs) should be followed at all times. Specific control measures for identified hazards are listed below.

Transmissible Spongiform Encephalopathies

All tissues from cattle with clinical BSE should be incinerated so that they are eliminated from all feed and food chains.

In countries where BSE has occurred, depending on its incidence, consideration should be given to placing restrictions on the use of meat and bone meal derived from specific bovine tissues in ruminant feeds. A similar consideration should be made in countries where a risk assessment indicates that the cattle population has been exposed to infection.

In countries where BSE and sheep scrapie have occurred, consideration should be given to placing restrictions on the use of ruminant derived protein from the feeds of ruminants.

In countries where BSE has not occurred, but where sheep scrapie is present, consideration should be given, dependent on the incidence of scrapie and the time/temperature processes used for the rendering of ovine carcasses and tissues, to placing restrictions on the feeding of ovine derived protein to ruminants.

The measures listed above may require re-evaluation in the light of future research findings on the inactivation of TSE agents during rendering.

Cross contamination of cattle feeds with meat and bone meal produced from the rendering of potentially infected cattle tissues should be prevented.

Biological Agents

Salmonella enterica, Toxoplasma gondii, Trichinella spiralis are sensitive to heat and are readily killed if the manufacture of feed involves pasteurisation.

Protocols developed for GMP must include measures which prevent recontamination of heat treated feed by these agents.

Veterinary Drugs

Only products licensed for administration to food producing animals should be used and the withholding time should be observed before milk or eggs are used for food or animals are sent for slaughter. Adherence to the Codex Code of Practice and Guidelines for Control of the Use of Veterinary Drugs and of Veterinary Drug Residues in Foods (CAC/RCP 38-1993) will ensure animal feeds do not contribute to excessive veterinary drug residues in foods.

Agricultural Chemicals

It is essential that the levels of agricultural chemicals in feed are sufficiently low that their concentration in food is consistently below the established maximum residue limits such as those limits established by Codex.

Mycotoxins

Feeds contaminated with mycotoxins in excess of established national maximum levels or international established maximum levels such as those established by Codex, should not be fed to animals producing milk, eggs or other tissues used for human consumption.

Grain and cereals should be stored under conditions of low moisture. Mould inhibitors can be added to reduce fungal growth.

Mycotoxin contaminated grains can be used for alternative purposes such as alcohol production, but by-products that result should not be fed to food producing animals.

Certain chemical substances and biological agents incorporated into feed, either intentionally or unintentionally, can result in hazards in food of animal origin and may enter feed at any stage of production up to the point of feeding.

The risks to human health associated with hazards involved in animal feeding are relatively low in comparison the risks arising from to foodborne hazards from other sources.

Where foodborne hazards originate in feed, the hazard should be adequately controlled.

Feed ingredients which do not pose any foodborne risk or for which any foodborne risk can be adequately controlled should not be prohibited from use in feed on the basis of food safety concerns.

Changes in feed or in the formulation of feed, as well as changes in feed processing methods, may result in changes in the risk from foodborne hazards which originate in feed. It is important that this be recognised and that potential risks be evaluated before any change is made.

The management of risk from foodborne hazards which originate in feed needs to be weighed against the potentially greater risks that would result from an inadequate or overly expensive food supply as well as the environmental risk that would result from the failure to recycle nutrients. There is a need for collaboration between all parties involved in feed and animal production, especially those in a position to provide veterinary clinical and epidemiological information, to establish the linkage between any identified or potential hazard and the level of risk. Such information is essential for the development and maintenance of appropriate risk management options and safe feeding practices.

Regulatory programmes should be established which ensure that foods of animal origin produced for human consumption are both safe and wholesome. In this context, the hazards which have been identified by the Consultation are well recognised, and suitable and appropriate control measures are in place in many countries. Examples include ante- and post-mortem inspection of slaughter stock, the control of the manufacture and use of veterinary drugs and agricultural chemicals, as well as residue monitoring programmes.

Though no conclusive evidence has yet been published, the Consultation determined it to be prudent not to exclude BSE as a potential foodborne hazard. It concluded that the risk that arises from this should be assessed and managed in exactly the same way as other foodborne hazards. This may necessitate the exclusion of certain material from feed for particular circumstances.

The disciplines that apply to international trade in both food and feed, as well as in feed ingredients were agreed to during the Uruguay Round of Multilateral Trade Negotiations and set out in the SPS Agreement.

The code of practice for good animal feeding that has been developed by the Consultation is intended to provide guidance which will minimise foodborne risks associated with feeds in a manner which is consistent with the principles of the SPS Agreement. The Consultation was of the view that adherence to this code should obviate the need for any trade restrictions on food or feed based on feed related human health concerns.

With respect to the production of food from aquatic animals using formulated feed, the food safety issues are the same as those applying to the production of food from terrestrial animals and no special issues apply. The Consultation concluded, however, that there may be food safety issues associated with the feeding of aquatic species through the fertilization of ponds by animal manure, agricultural by-products and other wastes.

It noted that this issue is to be considered by the WHO/FAO/NACA Study Group on Food Safety Issues on Food Safety Issues associated with Products from Aquaculture and concluded that was the appropriate forum for this issue to be addressed.

Recommendations

The Consultation made the following RECOMMENDATIONS.

1. The feed industry and the animal production industries should recognise their important role in the production of safe food and should evaluate the consequences to human health when using new feed ingredients, new suppliers or introducing new processing methods.
2. As quality assurance is applicable at all stages of food production to ensure the safety of the consumer, a code of practice for good animal feeding should be followed.
3. Manufacturers should provide adequate information to enable the quality and safety of feed to be maintained after delivery.
4. Known and potential risks to food safety should be re-evaluated as new information becomes available.

5. A code of good practice for the fertilization of ponds by the addition of animal manure, agricultural by-products and other wastes should be developed by the WHO/FAO/NACA Study Group on Food Safety Issues on Food Safety Issues associated with Products from Aquaculture and conveyed to the Codex Alimentarius Commission for possible inclusion in a Code of Practice for Good Animal Feeding.
6. The Codex Alimentarius Commission should consider for adoption the Draft Code of Practice for Good Animal Feeding.
7. The feed industry should assist developing countries by providing and promoting advice on good animal feeding practices.
8. FAO should support developing countries in the application of good animal feeding practices.

Chapter 6

Socio-economic and Biological Issues to Consider in Smallholder Poultry Development

Poultry provide major opportunities for increased protein production and incomes for smallholder farmers because of the following factors: small generation interval, the high rate of productivity, the ease with which poultry products can be supplied to different areas, the ease with which poultry products can be sold due to their relatively low economic values, minimal association of poultry with religious taboos, and the complementary role poultry play in relation to other crop-livestock activities.

Despite the potential of poultry production to attain rapid growth, constraints such as diseases, pests, predators, poor housing, poor nutrition, lack of knowledge about poultry production, lack of genetic improvement in commercially important traits in indigenous breeds, lack of capital and poor accessibility to markets can negatively affect poultry production in the smallholder farming sector. To boost income generation and nutrition for the smallholder farmers and their families these constraints should be addressed by carrying out on-farm research involving farmers, especially women who normally carry out the day to day activities of the poultry sub-system.

Since smallholder farming systems are bio-economically complex involving several kinds of resources and input/output flows, it would not be advisable to study poultry production only without considering other crop-livestock components of the farming system. Farming systems research (FSR) has emphasised traditional cropping systems, with little attention to livestock sub-systems, let alone poultry. The reason could be that the economic importance of poultry is not adequately appreciated by researchers and decision-makers because poultry products in the smallholder farming sector only passes through non-formal marketing channels. For the smallholder farmers to adopt new technologies meant to improve poultry production, the technologies should be compatible with local socio-cultural, political, economic and institutional conditions. This can only be done by carrying out more holistic FSR which go beyond biological production in crop and livestock sub-systems to include utilisation, marketing outputs and off-farm activities. Research methods should be more participatory than extractive.

This paper therefore highlights smallholder poultry research, development and production potential, constraints and challenges for the Southern African poultry researchers in the new millennium. The term 'smallholder farmers' refers to farmers in communal, resettlement and small-scale commercial areas.

Although poultry may be of low economic value in smallholder farming systems, it is an important source of protein in both rural and urban areas. Due to recurrent droughts in Southern Africa and decreasing grazing land, which tend to result in the reduction of the national cattle herds, poultry has the potential to become a major source of protein and a ready source of income in the smallholder farming sector. In Zimbabwe demand for day-old chicks increased from 10 million in 1988 to 24 million, while that of eggs increased from 11.5 million to 20 million in 1995 (Government of Zimbabwe 1995). While the 1963 to 1975 period saw an annual growth rate of livestock production in developing countries of 2.4 %, the annual growth rate for poultry production was much higher at 6.5 % (Hrabovszky 1981).

In most developing countries poultry production systems vary from being extensive to intensive. In Zimbabwe there are the large-scale poultry production and smallholder poultry production systems. Large scale poultry production units are characterised by huge capital investments, intensive management, mechanisation and specialisation. The smallholder poultry sector can be intensive characterised by exotic breeds of broilers and layers, semi-intensive or extensive dominated by indigenous poultry breeds that are not specifically classified. The indigenous poultry breeds in Zimbabwe are estimated to be between 15 and 30 million (Lambrou 1993). Although they are characterised by low production, indigenous birds are tolerant to diseases and are adapted to poor nutrition. The free-range poultry production system can best be described as a low input-low output system where the birds are rarely given any extra feed besides what they get from the surroundings. This system represents a part of a balanced farming system where little or no inputs are involved. The indigenous birds are also an important reservoir of genomes that may be used in the future to improve hybrid birds.

Of the 36 million day-old chicks produced annually in Zimbabwe, 50 % are purchased by the smallholder producers (Government of Zimbabwe 1995), but little information is available on what happens in this sector. This can be attributed to lack of support and development of rural poultry production by the government and development agencies. Some of the constraints to poultry production are diseases, pests, predators, poor housing, poor nutrition, and lack of genetic improvement in commercially important traits in indigenous breeds; lack of capital and poor accessibility to markets (Faranisi 1995; Tadelle 1996; Government of Zimbabwe 1995; Muchenje and Sibanda 1997). Improved management of these birds through, for example, supplementing indigenous birds with energy, protein and calcium can increase poultry productivity in extensive systems (Tadelle 1996).

Despite the constraints listed above, there are several factors that make expansion of poultry production in the smallholder farming sector attractive. Poultry meat and egg production enterprises grow very rapidly, not only because of the biological, physical and economic potential, but also because well-run poultry enterprises provide high levels of protein output per unit of feed (Hrabovszky 1981). Unlike pigs, poultry are associated with few religious or social taboos.

Supply of poultry products in poor countries is technically possible because the birds easily adapt to most areas of the world, have a low economic value, have a small generation interval and a high rate of productivity (Smith 1990). Meat from poultry can be produced within eight weeks and eggs within 18 weeks of the first chick being hatched. In addition, poultry play an important complementary role to other crop-livestock activities.

Poultry production is a dynamic sub-sector within livestock production systems and improvements in management and technology in smallholder poultry production systems will ultimately improve household income and nutrition. This can be achieved through participatory research in smallholder poultry production. Such research activities would result in poultry production models clearly describing the input-output relations in, for example, extensive poultry production systems and interrelationships between poultry production and other crop-livestock sub-systems.

Importance and Uses of Poultry

Poultry are an important source of ready income and protein in rural areas. Rural poultry also integrate very well and in a sustainable way into other farming activities (Tadelle 1996). Furthermore, rural poultry play important cultural and social roles in rural communities. The main objectives of rural poultry are producing eggs for hatching, sale and home consumption, and birds for sale, healing ceremonies, traditional offerings, replacement, home consumption and for gifts to visitors and relatives (Mutisi and Kusina 1996; Tadelle 1996; Muchenje and Sibanda 1997). Money from the sale of the birds is used to buy immediate household requirements such as food, dairy feeds and to pay school fees, among other purposes. By eating leftovers from the kitchen and insects such as cockroaches, birds perform a valuable sanitary function in villages. Poultry manure can be used as field manure or as feed supplement for ruminants.

Poultry can be kept for socio-cultural or financial purposes. Under such circumstances introducing management changes such as regular watering and feeding, cleaning the fowl-runs and taking care of the young chicks can be complex. When designing programmes to expand poultry production it is important to consider issues related to gender and labour-sharing within households. Some studies have shown that poultry keeping in most developing countries is the responsibility of women (Tadelle 1996; Muchenje and Sibanda 1997). Understanding the roles of each household member is important in sustainable poultry production.

Poultry Productivity in the Smallholder Farming Sector

Free-range poultry production levels are generally low and most flock sizes average 10 to 20 birds; two to three clutches per year, with an average of 10 to 15 eggs per clutch. The average annual egg production is between 30 and 60 small sized eggs (Smith 1990; Tadelle 1996; Muchenje and Sibanda 1997). The low production levels can also be attributed to high chick mortality levels (Tadelle 1996), diseases, pests, predation, poor housing and poor management (Muchenje and Sibanda 1997). According to Tadelle (1996), chick mortality, which may be over 60 % between hatching and end of brooding, represents a major loss in scavenging village chicken production systems.

This may be due to a combination of poor housing, competition for the same feed resource base with stronger and more vigorous members of the flock, low protein and energy of available feed, the low hatching weight of chicks, predation and diseases.

In addition to keeping indigenous chickens, some farmers in rural areas of Zimbabwe rear broilers and layers. In most cases, broilers are reared and targeted for the December, April and August public holidays when the market is lucrative.

The average broiler flock size in Zimbabwe is 25 birds, although some farmers can buy as many as 100 chicks at a time (Muchenje and Sibanda 1997). It is important to study the different production systems in the smallholder sector to have an insight into the problems and come up with programmes to alleviate such problems.

Besides chicken production, other poultry species such as turkey, pigeons and ducks are kept in the smallholder sector. These are mainly for family consumption.

The average turkey flock size in Nharira-Lancashire and Sanyati Smallholder Farming Areas of Zimbabwe was 2ýÿ5 birds, and most farmers were interested in expanding their flocks (Muchenje and Sibanda 1997). In the same areas pigeon flock sizes ranged from 2 to 52 birds. Little research work has been conducted on the production systems of these minor poultry species in most developing countries (Faure et al. 1996; Kusina and Muchenje 1997).

Non-adoption of Poultry Production Technology by Farmers

Despite considerable research and extension efforts in agricultural systems, smallholder farmers continue to shun most available technologies, particularly those that relate to livestock.

This may be attributed to inappropriate research methodology, lack of inter-disciplinarity within research teams, and weak farmer-research-extension interaction (Francis and Mutisi 1997). Some of the factors resulting in slow progress in smallholder farmers adopting new technologies in poultry production are highlighted.

Table 1: Comparison of characteristics of smallholder crop and poultry production systems.

Factor	Crops	Poultry	Implication for on-farm poultry research
Mobility	Stationary	Mobile	Difficult to measure and control non-experimental variables. High chances of missing data.
Life cycle synchronisation	All units synchronised	Units seldom synchronised	Difficult to find comparable units.
Multiple units	Only grain or tuber and residue	Many:- meat, manure, cultural	Difficult to measure and value treatment effect
Non-market inputs or outputs	Few	Many	Difficult to value input/output
PRODUCER ATTITUDES	Impersonal	Personal taboos	Difficult to cull or impose treatments or production "deemed" harmful to animals
Observation units	Many	Few	Large statistical variability:- experimental design should allow for loss of experimental unit
Variability of observations	Lower	HIGHER	Experimental design should allow for loss of experimental unit

Adapted from: Francis and Mutisi (1997).

Farmers are worried to expose their animals to risk through research or trial of new technologies. Therefore farmers usually prefer that their weakest and/or oldest animals be used in trials, hoping that they would be improved (Francis and Mutisi 1997). This results in researchers having difficulties in coming up with proper experimental designs. Furthermore, there is no synchrony in the life cycles of the birds, making it difficult to obtain enough birds which meet the specific criteria demanded by researchers, for example the same breed, laying stage, sex and age.

The implication on on-farm poultry research is that usually there are insufficient replications. Management of poultry by farmers varies considerably. This is mainly due to different physical and biological site characteristics, and different socio-economic conditions among farms. Such variations result in wide statistical variability, most on-farm trials being associated with large coefficients of variation and hence difficulties in interpreting results, and no statistically significant differences between treatments when such effects may exist. Incidences of missing data may be high since farmers may fail to avail their birds when measurements are taken. The complex nature of smallholder farm production systems makes it difficult to account for all the possible interactions and environmental effects. There is therefore a need to explore alternative methods of evaluating such trials. Analyses such as nested designs and incomplete block designs need to be considered in such circumstances.

Multiple non-cash outputs of poultry, such as manure, traditional purposes, home consumption and status compound the difficulty in analysing poultry enterprises. Since poultry products consumed by the farming family only passes through non-formal marketing channels, researchers and decision-makers do not adequately appreciate the economic importance of such products.

Researchers also usually regard the years when crop failure is experienced as unrepresentative and often stop data collection. This results in recording of peak cash output from poultry being missed. This calls for supplementing conventional economic analyses with the farmers' own assessment of the importance of poultry outputs in any research and development effort. Predetermined agendas and duplication of on-station methodologies should be avoided. The search for quantifiable data should be done without ignoring farmers' on-farm trial results based on their experience over the years. On-farm research should be participatory rather than being extractive.

While both biological and socio-economic aspects of smallholder farm production are important, there is lack of inter-disciplinarity in most on-farm research activities. Research teams should be fully integrated and inter-disciplinary involving both biological and social scientists (Hanyani-Mlambo 1997; Francis and Mutisi 1997). Studies should address the integrated nature of livestock and crop production subsystems and off-farm activities. There is also a need to link research and extension activities.

In most on-farm studies there is a weak farmer-research-extension interaction. Participation of resource-poor farmers is low yet technological recommendations tend to be general. Extensionists should also participate in research as opposed to only assisting in transferring technology from researchers to farmers (Francis and Mutisi 1997). There should be joint planning and review meetings, workshops and conferences between researchers, extensionists and farmers. Consideration of farmers' knowledge/technology and language use are pivotal in adopting any new technologies. Researchers should therefore view farmers as consultants, collaborators or colleagues. Participatory research should be introduced/improved in agricultural colleges' and universities' syllabi. Holistic farming systems research (FSR) that go beyond biological production in crop and livestock subsystems to include utilisation and marketing of outputs, and off-farm activities are needed.

Some impressive advances have already been realised through utilisation of the FSR approach. The International Livestock Research Institute (ILRI) has been applying the FSR approach mainly on pastoral systems (Hanyani-Mlambo 1997).

In Zimbabwe the FSR Unit in the Department of Research and Specialist Services (DR and SS) has carried out research using the FSR approach in Chivi and Mangwende Communal Areas. The Faculty of Agriculture, University of Zimbabwe has been using the same approach in studying crop-livestock farming systems in several smallholder farming areas in Zimbabwe.

Possible Changes to Improve Productivity of Indigenous Flocks

According to Smith (1990), the low input-output levels are economically efficient, since with the low output from the individual birds, the inputs are low or virtually non-existent. Small management changes, such as regular watering, night enclosures, discouraging hens from getting broody, vaccination against common diseases, small energy and protein supplements and caging chicks can bring about significant improvements in the productivity of indigenous birds (Tadelle 1996).

Smith (1990) estimated that under scavenging conditions the reproductive cycle consists of a 10-day laying phase, a 21-day incubation phase and finally a 56-day brooding period. This implies a theoretical maximum number of 4.2 clutches per hen each year, although in reality the number (Mutisi and Kusina 1996; Muchenje and Sibanda 1997). In the Eastern hills of Nepal, indigenous birds kept under semi-extensive management conditions produced 125 eggs in 11 clutches per annum, compared to 3ýÿ4 clutches per annum produced under normal scavenging systems (Smith 1990). By improving existing management, it is possible to increase indigenous poultry production in the smallholder sector. Benefits from genetic improvement can only accrue if existing management is improved.

Feed Resources and Requirements

The feed resource base for rural poultry production is scavenging, and consists of household waste, anything edible found in the environment and small amounts of grain supplements. It is not constant. The portion that comes as a grain supplement and from the environment varies with activities such as land preparation and sowing, harvesting, grain availability in the household and season and the life cycles of insects and other invertebrates.

According to Tadelle (1996), protein supply may be critical, particularly during the drier months, whereas energy supply may be critical during the rainy season. In the absence of an event that diminishes the flock biomass, such as diseases, the village flock will normally be at the maximum biomass that can be supported by the available feed resource base. Any additions to the flock that increase the biomass result in increased survival pressure and selection against the weakest members of the flock. The feed resource base is usually deficient in protein, energy and probably calcium for layer birds.

Determination of the nutritional status of indigenous birds and an assessment of changes in their diets throughout the year can be done by analysing crop contents and doing carcass measurements. After slaughtering the birds, materials in the crop are combined, mixed and physically identified (Feltwell and Fox 1978).

Seeds, insects, worms and plant materials are separated, weighed, and recorded, and the proportion of each calculated. The collected crop contents can also be analysed for dry matter (DM), crude protein (CP), ether extract (EE), crude fibre (CF), and ash according to AOAC (1985), for calcium (Ca) by atomic absorption spectrophotometre, and for Phosphorus (P). The metabolisable energy (ME) levels of the contents can also be determined by an indirect method, using the following equation (Wiseman 1987):

True ME (Kcal/kgDM) = 3951 + 54.4EE-88.7CF-40.8Ash.

Different parts of the carcass can be separated and measured. Carcass weight, gastro-intestinal tract (GIT), proventriculus, crop, gizzard, liver, ovary and oviduct weights can also be measured. Lengths of parts of the body such as the GIT, caecum and circumference of the upper leg (thigh) can also be measured. Carcass weight can be taken after the offal comprising feathers, lower leg, heart, crop, pancreas, lungs, head and digestive and urogenital tracts have been removed. Then the carcass dressing percentage can be calculated. The possibility of increasing free-range poultry production through energy and protein supplementation, and improved housing also needs to be investigated. For the benefit of those smallholder farmers who keep broilers and layers, the need to consider home grown feeds as substitutes is not overemphasised. This is against a background of ever escalating feed costs.

Breeding in Extensive Production Systems

Practically all village producers breed and rear indigenous day-old-chicks for themselves in order to maintain or increase the number of their flocks (Smith 1990; Lambrou 1993; Mutisi and Kusina 1996; Tadelle 1996; Kusina and Muchenje 1997; Muchenje 1997). The breeding of indigenous birds is not controlled because of the mixing of different flocks.

The breeding season for village birds varies according to the condition of the bird, the natural availability of feed and the weather. According to Lambrou (1993) and Kusina and Muchenje (1997), the breeding season for village birds in Zimbabwe is from September to April. A strategic supplementary feeding during the breeding season will improve the indigenous bird condition at breeding, the breeding efficiency and will increase egg production, fertility and hatchability. There is also a need to come up with a more effective and economic ratio of males to females in indigenous birds. Is the 1: 15 ratio generally recommended in Zimbabwe (Lambrou 1993) the most economic and effective?

Markets

Ways in which farmers can improve their image and offer a competitive and consistently quality product on a regular basis to their customers need to be explored. Furthermore, aspects such as transport, market accessibility, marketing channels and organisation need to be addressed if farmers are to benefit from research. In rural areas poultry products are usually sold locally. Densely populated areas such as growth points, service centres, boarding schools and other government institutions are important markets for poultry products.

In Zimbabwe broilers are usually sold when they are at least six weeks old, although some farmers keep the birds for more than twelve weeks, depending on season and demand. Where people are into broiler production, indigenous birds are rarely sold and cost less than broilers. However, in areas where broiler production is not common, indigenous birds are usually sold. Usually, there is no defined selling age for the indigenous birds. They are sold depending on supply and family financial demands, for example, when school fees are needed.

In a survey carried out in Nharira-Lancashire Smallholder Farming Area of Zimbabwe (Muchenje and Sibanda 1997), layers were reported to start producing eggs at between 24 to 28 weeks and the eggs were being sold at the same price, irrespective of their size. The market for broilers and eggs in all areas was said to be favourable during holidays such as Christmas and Easter. The market for the less common poultry species, such as turkeys, pigeons and ducks, is normally not clearly defined.

Other Issues to Consider in Smallholder Poultry Research and Development

Conservation of Genetic Resources in Poultry: Conservation of domestic animal diversity (CDAD) encompasses characterisation, identification, monitoring and utilisation to ensure management for best short term use and longer term ready availability (Moyo 1995). There is a need to conserve "adaptive" traits as opposed to "commercial" traits. The conserved genetic resources will be future sources of unique genes and will be useful when environmental concerns necessitate change in production systems. Indigenous birds in most African countries are not adequately characterised to understand the existing diversity to facilitate development of rational utilisation and conservation strategies. Biotechnology techniques such as gene mapping could be used to characterise indigenous birds. It should, however, be noted that several Southern African countries face a lack of infrastructure for breeding purposes and have relatively small defined populations (Wollny 1995a; Wollny 1995b). There is a need to consider characterisation of indigenous poultry breeds if genetic conservation for economic, scientific, cultural and social, development and sustainability (Wollny 1995b) is to succeed.

Computer Simulation Modelling in Farming Systems and Scientific Research

Computer simulation models are useful tools, which can be used in studying farming systems. They can be used to describe the behaviour of business, economic, social, biological, physical or chemical systems over periods of time. The tremendous revolution in computer technology has resulted in growth in simulation modelling and this has eased the integration of agricultural sciences.

Computer simulation models mimic real life situations. They can be used to make the same decisions that would be made if it were feasible to experiment with the system itself. Since a system is a collection of entities acting and interacting together towards accomplishment of a logical end, computer simulation models can play an important role in systems analysis. A model for semi-scavenging poultry rearing has been developed in Bangladesh and has been applied with considerable success to the benefit of rural, poor women (Rahman et al., undated). The model included the indigenous breed (desi) and its crosses with exotic breeds. Some of the traits, which were considered in the model, were age of first egg, eggs/hen/year, mortality rate, and mortality due to predator.

A computer simulation model requires formalisation of current knowledge in the form of equations, solutions of the equations and interpretation of the results. Models have been used in studying metabolism. They integrate a variety of nutrient inputs for their outputs as products, heat or other metabolic consequences and satisfy an objective of intake models- that of integrating a wide variety of factors known to influence intake. The interaction between physical and metabolic pathways affecting feed intake have therefore been successfully integrated by modelling.

A bio-economic model developed in Brazil has been used to help in identifying priority areas for applied research, to explore various combinations of herd potential and level of feeding, the behaviour of different herd sizes, and the potential returns from specific technical innovations (Brockington et al. 1983). Another model was developed to assess the risks and returns from establishing improved pasture in an extensive beef breeding enterprise, for use by farmers and extensionists (Beck et al. 1982).

Conceptual Structure

The main components of the input-output system of free-range poultry production systems and poultry-other-crop-livestock systems can be incorporated into the models. Concepts and quantitative values from literature and field studies can be used to develop the models. Quantitative data may be used to describe the relationships and define default values, and the qualitative data can help in coming up with reasonable assumptions for the model.

Model Building and Validation

The discipline of mathematics, which provides appropriate structure with which to represent the biological and socio-economic concepts, can be used to interrelate the component parts. The processes within the systems are normally represented by sets of equations. Laws of probability guard stochastic models. They involve use of random variables, for example, inheritance.

This means that a spread of results can be obtained from the same initial values and parameter values. Deterministic models give definitive predictions, without any probability distribution. An independent variable causes consistent changes in the dependent variable. Deterministic models can therefore be used for determination of nutritional requirements of free range chickens, and egg output for both free range chickens and intensively produced layers.

They can also be used to predict the flock biomass of birds the scavenging feed resource base can support. Stochastic models can be used for calculating aspects such as chick mortality rates. Mathematical models can also be used to simulate the behaviour of the poultry-other-crop-livestock systems free-range poultry production systems. Models that predict the effect of flock size on scavenging feed resource base, weights of chickens, mortality; output for sale/consumption and flock composition may be incorporated.

The effects of changes in fixed parameters on model performance need to be examined. This is done to test the correctness of the internal workings of the system and the model assumptions. Recommendations, as regards the use of the model and the possibility of redesigning it, can then be made.

The Future

Poultry are an important component of farming systems in smallholder farming areas. Research efforts in poultry should be targeted at aspects ranging from feeding, breeding, disease control and husbandry to obtain a better understanding of poultry production systems in relation to other crop-livestock systems and the changes in socio-economic conditions in smallholder set-ups. Biological factors are not the only factors considered by farmers when adopting technologies to suit their local farming circumstances. Instead, socio-economic factors are also important to the farmer and, therefore, should not be ignored. Both objectively measured variables and subjectively measured variables should be studied to produce more meaningful research results.

Detailed studies of most of the issues raised above would contribute to improved poultry production and contribute towards improving family income and nutrition. Qualitative data can be gathered in a once off/single visit.

Reliability of such data depends on the ability of farmers to recall past events. Smallholder farmers do not usually keep farm records! More reliable data are obtained through monitoring studies and discussions with farmers. All the different aspects of poultry covered in this paper should be incorporated into a model together, rather than individually, in order to identify combinations that optimise smallholder poultry production and improve smallholder household's income and nutrition.

Hormones in Animal Production

Hormone-dependent sex differences in growth rate have been known for a long time. It has also been known that growth rate and FCE (feed conversion efficiency) are higher in intact males than in castrates. It was natural, then, that the availability of hormones and other natural or synthetic substances displaying hormonal activity led to experiments aiming at their use to increase production.

Beginning in the mid-1950s, DES (diethylstilboestrol) and hexoestrol were administered to cattle increasingly in the US and the UK respectively, either as feed additives or as implants, and other types of substances also gradually became available. In general, such treatment has resulted in 10–15% increases in daily gains, similar improvements in FCE and improvement of carcass quality (increased lean/fat ratio). Thus there has been a substantial reduction in the amount of energy required per unit weight of protein produced (1,2), and the economic implications of this have been great.

While the use of hormonally active substances in animal production rose, opposition to their use also increased, because of the theoretical possibility that residues in edible tissues might endanger consumers. The factors leading to the ban on DES in the US, first imposed in 1973, have been described. Several reports confirm that DES endangers the health of animals and man, when repeatedly used in large doses. However, as regards risks due to the presence of residues in meat produced according to regulations, no documented deleterious effects have ever been reported in man, either from DES or any other substance with hormonal activity.

A distinction should be made between the hormones as such, for which the metabolism in the body is relatively well known, and synthetic or other substances for whose metabolic inactivation the body may not possess the enzymes necessary.

When natural hormones are used in animal production, claims of zero-tolerance residue levels are not meaningful, since these compounds occur in detectable and highly variable concentrations in body fluids as well as in the tissues of all animals, treated or not (6,7). For other substances with hormonal activity the situation is different. However, when residue levels are extremely low, it seems reasonable to weigh the potential risks against the undisputed positive effects some of these compounds have in animal protein production.

Hormone Preparations Used in Animal Production

Hormones of Endogenous Origin

These comprise the "classical" steroid sex hormones, oestradiol-17β, testosterone and progesterone. The two former are used either in the free form or as esters, mainly those of propionic or benzoic acid. Esterification generally causes prolongation of the half-life of the compounds in the body by 40 to 50%. The natural hormones having low bioavailability when administered orally, owing to rapid conjugation and metabolic transformation in the liver, they are therefore administered by subcutaneous implantation.

Of the *oestrogens*, the stilbene derivatives diethylstilboestrol (DES) and hexoestrol possess high biological activity and have been used most widely. They are active orally as well as by implantation. Other orally active oestrogens include ethynyl-oestradiol, a more slowly metabolized derivative of the true hormone, with higher activity. An oestrogen with an entirely different structure is zeranol, a derivative of a resorcylic acid lactone occurring in the fungus *Giberella zeae*.

The *synthetic androgens* comprise a large number of substances, most of which are steroids. Of these, trenbolone acetate (TBA) possesses strong anabolic properties and has received much attention during recent years, used alone or in combination with an oestrogen. Another anabolic steroid is methyl-testosterone.

Of *synthetic gestagens*, only one will be mentioned here: melengestrol acetate, which stimulates growth in heifers but not in steers, and which can also be used for the suppression of oestrus. Numerous other gestagens also exist, but at present few other than progesterone and melengestrol acetate are used to stimulate growth.

In addition to these substances, numerous others exist, and some of them are used more or less frequently in clinical veterinary medicine. However, clinical applications of hormones are not considered to be of consequence to the consumer, since such treatment is much less frequent than the use of hormones to promote growth.

Hormone preparations in current use as growth stimulants, which also shows modes of application, dosages, etc. It will be noted that almost all preparations currently in use are based on implantation, the site usually being the base of the ear, or less frequently, the dewlap.

Range of Application

In cattle the use of hormones is limited to veal calves and beef cattle. Veal calves are produced mainly in continental Europe, to an extent of about 8 million per year. Research has demonstrated that hormone treatment improves growth rate, nitrogen retention and FCE during the five- to six-week period before slaughter. Beef cattle, including steers as well as heifers, were treated in large numbers, especially in the USA and the UK, with DES or hexoestrol, administered orally, until the use of these compounds was restricted. During the last several years, practice has changed dramatically in the direction of increased use of implants of natural steroids, synthetic anabolic steroids and the phyto-oestrogen zeranol.

In *sheep*, especially in wether lambs, some increase in gain has been reported, but results are somewhat ambiguous. In *swine*, hormone treatment may increase growth rate, FCE and lean/fat ratio of the carcass in male castrates.

Poultry generally do not appear to respond to oestrogens by increased gain but by changes in lipid deposition. In male and female turkeys, androgens have recently been reported to increase growth rate as well as FCE.

Modes of Application

When DES was used as a feed additive, a usual procedure was to start treatment of steers at a body weight of 360 kg and continue administration for 120 to 170 days. Since restrictions on its use were imposed, most preparations have been administered as implants, whose effect is usually limited to 80 to 100 days. Practice varies with management systems.

Table 2: Hormonally-active substances used in animal production

Substances	Dose levels	Form	Main use - Animals	Trade name
Oestrogens alone:				
DES	10–20 mg/day	feed additive	steers, heifers	
DES	30–60 mg/day	implant	steers	
DES		oil solution	veal calves	
Hexoestrol	12–60 mg	implant	steers, sheep, calves, poultry	
Zeranol	12–36 mg	implant	steers, sheep	Ralgro
Gestagens alone:				
Melengestrol acetate	0.25–0.50 mg/day	heifers		
Androgens alone:				
TBA	300 mg	implant	heifers, culled cows	Finaplix
Combined preparations:				
DES + Testosterone	25 mg 120 mg	implant	calves	Rapigain
DES + Methyl-testosterone		feed additive	swine	Maxymin
Hexoestrol + TBA	30–45 mg 300 mg	implant	steers	
Zeranol + TBA	36 mg 300 mg	implant	steers	
Oestradiol-17β + TBA	20 mg 140 mg	implant	bulls, steers calves, sheep	Revalor
Oestradiol-17β benzoate + Testosterone propionate	20 mg 200 mg	implant	heifers, calves	(Synovex H (Implix BF
Oestradiol-17β benzoate + Progesterone	20 mg 200 mg	implant	steers	(Synovex S (Implix BM

Animals may be implanted at live weights from 270 to 450 kg. Depending upon the age and weight at the time of implantation, the animals are either slaughtered at the end of this first period, or fed for an additional period, either without further treatment or after a second implant to act for another 80 to 100 days.

Most types of implants in use are not removable, but removable types have recently been tested and their effects described. When tested in steers, no reduction in performance was recorded when the implants were withdrawn 32 and 39 days before slaughter.

Implantation is subcutaneous, usually at the base of the ear, thus eliminating the risk that residues of the implantation site will be present in edible tissue.

Effects of Hormones

Veal Calves

In veal calves, hormone treatment may begin at a body weight of about 65 kg, the animals being slaughtered at about 170 kg. Implants of 20 mg oestradiol-17β + 200 mg progesterone in males and 20 mg oestradiol-17β + 200 mg testosterone in females resulted in a 20% increase in daily gain and 21% higher nitrogen retention in the period studied.

In other studies, improvements of 10 to 12% in gain and 10% in FCE have been reported. Nitrogen retention is about 70% in the very young veal calf, but decreases gradually to below 40% at the age of about 15 weeks. For ages of 10 to 15 weeks, the average conversion of feed protein to body protein is about 40%; this rate can be increased to about 60% by hormone treatment.

The effective preparations were DES, oestradiol-17β, and the combination of TBA + oestradiol-17β. More recently, positive effects have been reported for zeranol alone (36 mg) and for zeranol (36 mg) + TBA (140 mg), with increases in nitrogen retention of the same order as for DES and E_2 + TBA. When zeranol + TBA was implanted at the age of 56 days, the growth rate up to day 106 increased by 18%.

Steers

The most extensive studies of the effects of hormones on growth and FCE have been carried out on steers, under strictly controlled conditions as well as in the field. Since 1975, most studies have involved implants of oestrogens alone, androgens alone, or combined oestrogen/androgen preparations, although many trials have also been based on oestrogen/progesterone combinations during recent years.

Oestrogen implants have included DES, hexoestrol, oestradiol-17β and zeranol. *DES implants* have, as in previous studies, resulted in an increase of about 12% in gain and in improvement in FCE of the order of 10%. *Hexoestrol implants,* usually in doses of 30 to 60 mg, have been shown in numerous experiments to lead to considerable improvement in growth rate and FCE.

In 19 trials carried out over the years on experimental husbandry farms in the UK, the overall average increase in gain produced by 45 or 60 mg hexoestrol implants was 0.16 kg a day, and in only 2 of the trials was it less than half that figure. *Oestradiol - 17β implants* alone (30 mg) have resulted in a 24% increase in gain and a 13% improvement in FCE. *Zeranol implants*, usually at 36 mg, have consistently improved gain as well as FCE. In a series of 21 UK trials over several years, the average response to zeranol implants alone was an increase in daily gain of 0.15 kg. Only in one trial was there no response. Similar results have been obtained in Ireland. Positive effects on gain in steers have been observed under a variety of experimental conditions, under controlled feeding, on *ad lib* feeding of standardized rations, and on pasture.

TBA implants administered alone at a dose of 300 mg have also had positive effects on growth, even if combination with an oestrogen has yielded better responses (*vide infra*). In a series of 8 trials in the UK, the average additional daily gain amounted to 0.09 kg, with considerable variation among trials. Similar results have been reported from Ireland.

Combined preparations. A number of trials have been carried out with implants containing two hormones. The combination of an oestrogen with an anabolic steroid, or with progesterone, has met with the greatest responses. *Synovex-S* has consistently increased gain as well as FCE, with responses averaging about 20% and 17% respectively. *Hexoestrol + TBA* (usually 30 or 45 mg hexoestrol + 300 mg TBA) has resulted in marked increases in gain, of the order of 30% and in FCE of the order of 20%. *Oestradiol-17+ TBA* (20/140 mg) has given similar results, as has *Zeranol + TBA* (36/300 mg), also recently tested.

Hormone preparations have also been tested in combination with substances such as monensin, which increase FCE by promoting propionic acid formation in the rumen. Results have varied from no effect to marked additional gain.

Reimplantation, tested under various forms of management with varying results, has not gained general acceptance. Lamming has stated that "repeat implantation of hormone is not likely to produce the benefits obtained from its initial use, since a second dramatic change in the endocrine balance of the animal is not likely to occur.

In addition, double implantation increases the possibility of exceeding the optimum dose rate and the chance of deleterious side effects occurring."

The evidence for highly significant positive effects on the growth rate and FCE of steers is thus beyond dispute, the most marked effects being provoked by implants combining an oestrogen with an androgen of high anabolic activity.

Bulls

Since the entire male animal produces its own anabolic androgen, testosterone, an effect of additional hormones similar to that for steers is not to be expected. The number of trials with bulls is also limited. Positive effects on gain have been reported using DES alone (2, 58, 59) and combined oestrogen/TBA implants; in other studies, no effect on gain has been recorded, while a certain increase in the deposition of fat in the carcass has been observed.

Heifers

Recent trials with beef-producing heifers have mostly been based on the use of an androgen, although oestrogens have been tested, alone or in combination. Thus, zeranol has been reported to increase gain, while in other trials no response has been observed. TBA administered alone (300 mg) has led to increases in weight gain and FCE of the order of 36% and 25% respectively. In other trials, combinations of an oestrogen with TBA or testosterone have yielded significant growth responses. In general, it appears that the effect of TBA alone in heifers corresponds closely to the effect of combined oestrogen/TBA implants in steers.

Sheep

Trials have mainly concerned wether lambs, and positive effects of hormonal treatment have been reported using DES, hexoestrol and zeranol, although other reports have indicated that zeranol yields no significant effects. Wether lambs implanted with TBA + oestradiol-17β have shown increases in gain, carcass weight and FCE. In general, however, the results obtained in sheep thus far do not warrant the same clear-cut conclusions as for steers and heifers.

Swine and Poultry

There is little evidence that existing hormonal preparations influence the growth rate and FCE to an extent that would be interesting from a practical point of view. The lean/fat ratio in male castrate and female pig carcasses may be increased by the use of oestrogen/androgen combinations. In poultry, redistribution of fat in the body is a known effect of oestrogens. Recent research indicates improved growth rate and FCE using androgens in young male and female turkeys.

Undesirable Side Effects in Treated Animals

Reported side effects of hormone treatment for growth stimulation are few and generally concern the use of oestrogens in steers. Changes in body conformation such as feminization and raised tail-heads were described as early as 1958. Similarly, bulling has occurred with increased frequency, although in most animals it is limited to the first few days after implantation. However, it has been reported from Kansas that 2.2% of all steers fed in pens have to be removed, at an estimated loss of $23 per head.

In a study of the effect of reimplantation of oestrogens in steers, all animals were given a 30 mg DES implant at a live weight of 260 kg, and then reimplanted 91 days later, with either 30 mg DES or Synovex S. Following the second implant, the frequency of the steer-buller syndrome was 1.65% for the DES-DES group, and 3.36% for the DES-Synovex S group. The economic advantage of using DES + DES was estimated at $1.15 per head. The steer-buller syndrome is a special problem in feedlots.

Mechanism of Action of Hormones

No reliable explanation of how the growth-promoting hormones act has yet been furnished. Some observations indicate an indirect influence through changes in the balance of endogenous hormones. Thus there have been reports of DES and TBA increasing the levels of growth hormone and/or of insulin in plasma; these hormones are known to stimulate amino acid transport across the cell membrane.

However, others have found no such effect. Bulls fed DES (10 mg/day) over two years had significantly higher plasma testosterone levels than controls; those levels are positively correlated with growth. Recent experiments indicate that DES reduces the rate of muscle catabolism in steers.

As regards the anabolic androgens, evidence exists indicating competition with glucocorticoids for receptor sites on the muscle cell membrane. Since glucocorticoids have a catabolic effect on tissues, their displacement from muscle cells would reduce catabolism. TBA alone, and even more when combined with oestradiol-17β, causes a marked decrease in the concentration of total thyroxin in plasma of steers (82). In another study, combined oestradiol-17β-progesterone implants (20 + 200 mg) in steers caused a uniform but slight increase in thyroxine binding capacity. The significance of these findings is not yet clear.

Levels of Endogenous Hormones in Body Fluids and Tissues

Any discussion of possible health hazards connected with the use of hormones in animal production must take into account the normal occurrence of hormones and their metabolites in body fluids and tissues, and the fact that the levels of these hormones vary greatly, according to the physiological state of the animal.

Table 3: Concentrations of endogenous hormones in edible tissues of farm animals

Animal/tissues	Oestrone pg/ml	Oestradiol-17β pg/ml	Testosterone pg/g	Progesterone pg/g
Veal calf				6
muscle		< 100	70	
liver		< 100	47	
kidney		< 100	685	
fat		< 100	340	
Bull				
muscle			335	
liver			749	
kidney			2 783	
fat			10 950	
Heifer	20–40			16
muscle		12–13	92	
liver		38–71	193	
kidney		40–71	595	
fat		6	250	
Cow, pregnant				336
muscle		370–860		
fat	3 870	2 500–5 500		
Steers				
muscle	6	14		
liver	20	14		
fat	23	10		
Wheat germ oil	4 000 pg/g DES equivalent			
Soy-bean oil	2 000 000 pg/g DES equivalent			

Thus, oestrogen levels in the blood of female farm animals may vary from a few pg up to 5–6 000 pg per ml plasma. As to males, the plasma of stallions and entire male pigs contains high levels of oestrogens, although mainly in the conjugated form. Milk also contains oestrogens in very high concentrations in the first drawings after parturition; in non-pregnant animals, levels in the range of 80–100 pg/ml have been reported. More recently, reliable data have also become available concerning concentrations in edible tissues; some of these are presented in table below. For the sake of comparison, levels of oestrogen activity normally present in products of plant origin widely used in human nutrition are included.

Metabolism, Routes and Rates of Elimination

The general patterns of metabolism and elimination of endogenous hormones in farm animals have been outlined. In ruminants, testosterone and oestradiol-17β are rapidly converted to their epimers, biologically much less active, epitestosterone and oestradiol-17β. Progesterone is partially converted to androgens before excretion. In the pig, epimerization of testosterone and oestradiol-17β does not appear to take place to a significant degree. The faecal route of elimination dominates in ruminants, while in the pig urinary excretion is more important.

Progesterone

After repeated injections of progesterone to cows and steers over 2 to 3 weeks followed by C-progesterone for 2 to 5 days, the animals were slaughtered 2 to 3 hours after the last injections. Activity levels were 2 to 7 times higher in the fat, 3 times higher in the kidneys, and 13 times higher in the liver than in the muscle.

Excretion of radioactivity amounted to 50% and 12% in faeces and 2.0% and 1.2% in urine in cows and steers respectively. About 50% of the activity in muscle and milk was associated with unchanged progesterone, most of the remaining activity being associated with a mono-hydroxy compound. Cooking or frozen storage did not affect the nature or quantity of metabolites.

Oestradiol-17β

Following daily injections of 1 mg oestradiol-17β or its benzoate to heifers and steers for 11 days, followed by the C-compounds on days 12, 13 and 14, the animals were slaughtered 3 hours after the last injections, when residual levels were maximal. In muscle extracts, oestradiol-17β represented the major fraction of extracted activity (38 to 71%), followed by oestrone (17 to 45%). Levels in muscle were 161 to 225 pg/g and 40 to 86 pg/g for oestradiol-17 and oestrone respectively. In fat the levels were 3 to 5 times higher. The authors conclude that residual levels are extremely low when these hormones are administered as growth stimulants to growing/finishing cattle.

Glucosides of the 17β- and the 17β- epimers, and the glucoronide of the 17β- epimer are the major metabolites in cattle. When oestradiol-17β was administered orally to swine, plasma concentrations were very high 7 min after administration. Oestradiol was completely conjugated during absorption and its first passage through the liver. Some conversion to oestrone took place.

DES

The metabolism of DES in food-producing animals has been reviewed recently. The substance seems to be eliminated to a large extent in unaltered form. After oral administration of ^{14}C-DES to beef cattle, 99.5% of the radioactivity was excreted within 5 days after withdrawal. In liver extracts, radioactivity associated with DES-conjugate and free DES was found to be 75% and 25% respectively. Higher than background levels of activity were observed after withdrawal in kidney, liver, bile and urine/faeces for up to 5, 7, 9 and 11 to 12 days respectively.

The fate of 24-mg DES implants containing ^{14}C-DES and implanted in the dewlap of calves was studied over 98 days. Free radioactivity was almost completely associated with unchanged DES. At the time of slaughter, levels were less than 0.1 ppb in muscle and fat, and 1 to 1.5 ppb in liver and kidney. In a study in steers implanted with ^{14}C-DES, on day 120 after implantation radioactivity in muscle was not distinguishable from background.

It was above background in spleen, lung, adrenal glands and kidney, but less than levels corresponding to 0.5 ppb. In a similar study on steers, 120 days after implantation, levels in liver, kidney, lungs and salivary glands were in the range of 0.07 to 0.13 ppb of DES equivalent.

In a recent study of DES metabolism in rhesus monkeys and chimpanzees, most of the substance was excreted with the urine. Extracts in the organic and aqueous phase mostly contained unchanged DES in the free and conjugated form respectively. Current evidence indicates that the oxidative metabolism of DES leads to at least three compounds that may have cytotoxic or mutagenic activity, but these have not been identified as DES metabolites in ruminants, but in the mouse.

Zeranol

Using a gas chromatographic method with a sensitivity limit of 20 ppb, no residues of zeranol could be detected in edible tissue from cattle slaughtered 65 days following implantation of 36 mg, or from lambs 40 days following implantation of 12 mg. In another study, tritiated zeranol was implanted in cattle as part of 36-mg doses. Skeletal muscle obtained 10, 30 and 50 days following implantation contained no detectable residual activity. This confirms previous results based on the use of ^{14}C-labelled zeranol.

Trenbolone Acetate (TBA)

Trenbolone is a 17β-OH steroid esterified in the 17 position with acetic acid. Upon release in the organism the ester is rapidly hydrolyzed to the free compound TB-17-OH and acetate. In cattle the 17β-OH compound is rapidly transformed to its 17α-OH epimer, in the same manner as oestradiol-17β in this species. The 17α epimer possesses only about 5% of the biological activity of the 17β epimer. Another metabolite of TBA in cattle is the 17-keto compound, analogous to oestrone; quantitatively it appears to be of very little importance. Following intravenous injection of TBA, levels of TB-17β-OH and TB-17α-OH of 0.05 and 0.005, 0.10 and 1.0, 0 and 191 ppb have been recorded for muscle, liver and bile respectively. Other metabolites occurred in extremely small quantities in cattle.

Similar findings have been reported in studies based on the use of implants. The major route of excretion is by faeces. Metabolism studies of TBA thus clearly show that the substance is rapidly subjected to biological inactivation in cattle, mainly by epimerization of the free steroid to the 17α-compound, and that the major route of excretion is via the bile.

Residues in Edible Tissues of Hormone-treated Animals

Much work has been devoted to the development of sensitive methods of detecting hormone residues in meat from hormone-treated animals. As regards compounds given orally, it should in principle be possible to realize claims of zero-tolerance residue levels, by selecting the proper withdrawal time. During recent years, the use of implants has, however, gained in importance. While removable implants have been tested in steers, with no decrease in performance when withdrawn 32 and 39 days before slaughter, the wide use of non-removable implants makes residue studies important.

Determination of normal levels of endogenously produced natural hormones is also important, to enable risk evaluation to be carried out in realistic terms.

Several residue studies have been made of synthetic as well as natural compounds, mainly in cattle. When regulations governing dose, sites of implantation and timing in relation to slaughter are adhered to, residue levels of DES, hexoestrol and oestradiol-17β (106, 107) in edible tissues have generally been in the lower ppb to the ppt range, i.e. from a few ng/g down to some hundred pg/g of tissue. In the latter case there was almost complete overlap between values for untreated and treated steers after 105 days. Zeranol implants have so far not left detectable residues in edible tissue.

Most studies of androgens have concentrated on TBA. The ester being rapidly hydrolyzed, measurements of residues have been limited to the free compound and/or its major metabolite. Results based on radio-immunoassay of extracts or on radioactivity measurements have indicated levels in edible tissue of the order of 1 ppb or below. In a recent study using implants containing tritiated TBA in heifers, it was found that when slaughter took place 60 days after implantation, the major proportion of tritium-containing residues was not extractable with organic solvents.

In muscle 95.5%, in liver 94.4%, in kidney 98.8% and in fat 59.1% of the radioactivity remained in the aqueous phase, not quantifiable by radio-immunoassay. This suggests that the major part of the residues after TBA implantation occurs in a non-extractable, possible covalently bound form in tissues.

Residue levels of gestagens have been also measured, in connection with their use as growth stimulants. Residues of melengestrol acetate used as a feed additive in daily doses of 0.25 to 0.50 mg per head have consistently been below the sensitivity levels of the methods used (i.e., below 10 ppb in fat, liver, muscle and kidney), whether or not the compound was withdrawn 48 hrs before slaughter.

Hormones in Food: Meat from Hormone-treated Animals Versus Other Sources

According to the Agricultural Research Service, United States Department of Agriculture (ARS), the average per caput consumption of beef is 157 g per day in the US. Calculations show that 157 g of beef from an animal implanted 61 days before slaughter with a combined implant containing 20 mg oestradiol-17β + 200 mg progesterone or testosterone will contain 3.43 ng oestrogen and 19.5 ng progesterone or 16 ng testosterone.

Data on normal levels of these hormones in certain dairy foods, shows that some foods represent hormone sources vastly richer than meat from hormone-treated animals. Based on these values, and averages for consumption of various foods, the relative contribution of meat from hormone-treated animals to the total consumption of hormones has been calculated on the assumption of proper use of the hormones.

It is clear that in most cases the contribution from meat of treated animals is insignificant when hormones have been properly used, and must be considered to be biologically without impact. This becomes even more evident when seen in relation to normal endogenous hormone production in man, as illustrated in table below. It will be seen that even for oestrogens, the hormones considered the greatest risk, the maximal contribution from meat (assuming proper use of the hormones) is less than 0.01% in the prepubertal boy who represents the lowest endogenous oestrogen production.

Table 4: Hormones in certain dairy foods

	Oestrogens	*Progesterone*
	(pg/ml)	*(ng/ml)*
Milk, from non-pregnant cows	80	9.5
Milk, from pregnant cows	126	
Cream		73
Butter		133

Source: 103.

Thus far the discussion has been limited to the natural hormones. For synthetic substances the situation may be different. But again, considering the very low residue levels found *when* hormones have been properly used, the question may be raised whether the risk to the consumers is being grossly overestimated.

Table 5: Relative contribution of meat from hormone-treated steers to total hormone intake via food (per cent)

	Oestrogens	*Progesterone*
Child under 1 year	0.22	0.014
Child 6 to 8 years	1.56	0.1
Adult male	7.69	0.5

Source: Condensed from 103.

Table 6: Contribution of hormones from hormone-treated steers relative to total daily hormone production in man (per cent)

	Oestrogens	*Progesterone*	*Testosterone*
Prepuberal girls	0.00636	0.00078	0.005
Prepuberal boys	0.00826	0.00130	0.00244
Women			
Follicular phase	0.00018	0.00047	
Luteal phase	0.00007	0.01	0.0004
Men	0.00025	0.00048	0.00003

[1] The figures represent effective fractions (i.e. 10% of real fractions), to take into account the low bio-availability of the hormones absorbed orally.

Economic Implications of the Use of Hormones in Animal Production

In the production of meat for human consumption, a hormonally-induced increase in growth rate of the order of 10% evidently has major economic implications.

The improvement in FCE which usually accompanies the increase in gain adds to the economic benefits, and at the same time makes possible greater production of edible protein per unit energy used, and this in itself is of importance in a world lacking in protein supplies.

Some of the hormones that have become available recently appear on average to increase gain as well as FCE considerably beyond the 10% level, and in examining whether they should be approved for use in animal production, the risk/benefit analysis must take this fact into account.

Few analyses of the economic advantages of using hormones as growth stimulants appear to have been made. For the UK, a recent calculation is based on the estimated increased return to producers for 1 350 000 cattle treated over a 12-month period.

Table 7: Estimated increased return to producers from the use of hormones in animal production (12 months)

	Ralgro	Finaplix
1. Number of animals treated[1]		
Steers	675 000	480 000
Heifers	75 000	120 000
2. Average increase in daily gain (kg)[2]		
Steers	0.11	0.06
Heifers	0.05	0.06
3. Average increase in slaughter weight (kg)[2]		
Steers	10	5
Heifers	6	3
4. Estimated total increase in slaughter weight (carcase weight) (tonnes)	5 478	
5. Estimated overall increased gross margin per head to the producer (?)[3]		
Steers	21.60	10.25
Heifers	10.40	8.55
6. Estimated gross return (?)		
Steers	14 580 000	4 920 000
Heifers	780 000	1 026 000
	15 360 000	5 946 000
Total	21 306 000	
7. Estimated price per dose (?)	1.20	1.80
Total (?)	900 000	1 080 000
8. Net return (?)[4]	14 460 000	4 866 000
Total (?)[4]	19 326 000	

[1] Estimated from sales of the preparations during a year.

[2] Based on results from Meat and Livestock Commission trials using yard finishing cattle receiving only one implant.

[3] Based on 1978 data showing that 0.1 kg increase in daily gain gave an increase in gross margin of £13 per head, and that increase in slaughter weight averaged 85 p per head. From these figures are subtracted the cost of treatment.

[4] Does not include costs of veterinary services, etc.

Source: 111

Assuming that 1 155 000 of these were steers and 195 000 were heifers, and that the estimated daily gain was only 0.06 to 0.11 kg for steers and 0.05 to 0.06 kg for heifers, depending on the preparation used, the overall gross increased return was calculated at £21 306 000, without taking into consideration improvements in FCE.

These calculations must be taken as an example only. Availability of the various feeds, variations in feed and product prices as well as in types of management from time to time and from place to place may play an important role.

However, shortening the time required for producing a certain weight at slaughter will represent an economic advantage, especially under feedlot conditions, since non-feed costs also contribute significantly to the total cost of production (10 to 18 cents per head per day in the USA).

Alternatives to the Use of Hormones

Growth rates are influenced by many factors, especially genetic constitution and feeding. Over time, selection as well as improvements in management systems, feed composition and feeding programmes have contributed much to increasing productivity in meat as well as milk. Although it is difficult to evaluate the exact relative contributions of these factors, the overall improvements have been dramatic.

An example is the increase in milk yield per head in US dairy cattle. In the period 1944–1975, the number of dairy cows decreased by 33%, while the average yield per cow increased by 60%. These gains represented a saving of about 23 billion kg of total digestible nitrogen per year, the volume of milk produced remaining relatively constant. The saving is equivalent to about 1.1 billion bushels of maize. Data illustrating progress in beef production over the years are scarce, but increases in productivity similar to those for milk production are unlikely.

In addition to the use of hormones, many avenues are still open for increasing productivity in meat and milk production (see 115), including breeding programmes, regulation of rumen fermentation, optimalization of the balance between the indirect and direct feeding of the ruminant organism proper, and disease control.

Breeding Programmes

Systematic selection of high-quality sires, combined with an increase in the number of offspring from high-yielding females through embryo transfer, may bring about further improvements in beef and milk production.

In many countries, development along these lines has hardly begun. However, the establishment of effective breeding associations and the strict organization of programme planning and execution are prerequisites for realizing the potentials in this sector.

Regulation of Rumen Fermentation

The microbial systems in the rumen are extremely complex, and the balance between the various strains of bacteria is susceptible to changes brought about by many factors. The recent introduction of substances such as monensin offers great promise in altering the fermentation pattern to the benefit of productivity by increasing FCE.

Since the very extensive breakdown of carbohydrates and protein represents loss of much energy, research is currently being conducted in many laboratories in order to find new methods of increasing FCE.

Optimalization of the Balance between the Indirect and the Direct Feeding of the Ruminant Organism Proper

To a large extent, feeding a ruminant means feeding the rumen microbes which then themselves serve as feed for the organism proper. This is indirect feeding, expensive in energy. On the other hand, the ruminant possesses, in the postruminal part of its digestive tract, all the enzymes necessary for utilizing all types of nutrients except cellulose. The rumen microbes are necessary for the utilization of cellulose, which globally represents an enormous source of energy. However, it is possible to sustain an adequate microbial population in the rumen even when ruminal breakdown of part of the easily digestible nutrients is prevented.

Enabling nutrients to bypass the rumen will increase the utilization of feed for production, and also create a more adequate supply of amino acids. Increased rumen bypass of nutrients can currently be brought about by several means, including formaldehyde and heat treatment of protein-rich feeds.

A third method, aiming more at specific substances that may be rate-limiting for production (e.g. certain amino acids), or of significance in treatment of diseases, is protection against rumen degradation by such means as incorporation into the ration of long-chain fatty acid mixtures in the form of small pellets.

In the future, new methods of increasing rumen bypass will undoubtedly contribute significantly to increased productivity of ruminants.

Disease Control

Whatever management system is adopted, effective disease control is essential for productivity. In many areas of the world, infectious and parasitic diseases inflict heavy losses on animal production. A recent study has disclosed nearly a one-to-one relationship between investment in agricultural research and annual productivity of edible protein in ruminants.

An increase of about 45% in scientist/man years and a corresponding increase in funding for research and development is considered sufficient to raise productivity in this sector by 50%. Investment in disease control is an important aspect of this work. Annual world mortality losses from disease exceed 50 million cattle and buffalo, and 100 million sheep and goats. Non-lethal diseases are believed to lead to an equivalent reduction in production. Thus, investment in disease control holds great promise for future augmentation of animal protein production.

In these perspectives, the significance of hormones in animal production may seem marginal, leading to the question of what priority to give to the various efforts to increase productivity and production. In the global context it is, however, at least at present, impossible to adopt one approach to the exclusion of others. As long as preparations exist that combine positive effects on yield and feed utilization with low or non-existing risk to the consumer, there will be a market for them. What is more, the use of hormonally-active substances in the future may not be limited to those currently available.

Common to the present compounds, natural or synthetic, is that they are degraded in the body only to a limited extent. An entirely different situation exists for proteid hormones, which are broken down completely to amino acids, leaving no residues whatever. An example is the growth hormone which not only stimulates growth but also milk secretion, even in high-yielding cows.

This anabolic hormone is currently available only in small quantities for research. However, a recent breakthrough in the use of recombinant DNA technique has made large-scale microbial production of species-specific peptide hormones a realistic possibility. Combined with the development of miniaturized automatic delivery systems for subcutaneous use, a new era may be visualized as regards the use of hormones in animal production.

Methods of Measuring Hormone Levels in Animal Products

The question of the use in animal production of anabolic agents, including the natural steroid hormones and their derivatives, has been — and continues to be — badly stated, and therefore misunderstood. Those who brought the problem before the public had neither the information nor the technical competence, especially in biology, to speak with perfect objectivity.

Hormones are found naturally in many animal products (6, 7) and, as phytohormones, in many plants (3, 5). Only some of them raise public health problems, and these should be distinguished from the others. Indeed, a study of the practical and health aspects of the use of hormones in animal production should underlie all regulation of the matter, and therefore any decision on the measurement of hormone levels.

A considerable volume of documentation exists on the problem (6), which is as important for the developed as for the developing countries. Items 1 to 11 in the Bibliography of this paper indicate a number of general studies.

The Use of Hormones in Animal Production, and its Consequences

The first type includes those not found in mammals and birds but obtained exclusively through organic synthesis: diethylstilboestrol (DES) or stilboestrol, dienoestrol, hexoestrol and their derivatives. These compounds, active when administered orally or as implants, are harmful for laboratory animals, whether given directly or as residues in the flesh and offal of treated animals.

The use of these substances, a potential danger to human health, is prohibited in many countries, particularly since the methods of measuring DES levels in meat are unreliable, so that measurements must be made on the excreta.

The second type of anabolic agents consists of the natural steroids, normally present in all animals: testosterone, oestradiol, oestrone, progesterone. Secreted by the endocrine glands, these substances and their derivatives, although absorbed by the intestines, are deactivated in the liver. They produce measurable effects only when implanted.

The third type, found in the higher plants and moulds, consists of substances with a more or less clear oestrogenic and, at times, anabolic activity.

Among these phytoestrogens, only zearalenone has been isolated for use in animal production: as zeranol, it gives good results, but has an effect on genital formation and the level of thyroxine in plasma. It should be added that Sharaf and Gomaa attribute oestrogenic effects to Vitamins E, B_6, C and perhaps A, while Nelson et al. have studied the same effects in O-p'-DDT.

First mention should be made of cases of cancer of the vagina and hypogonadism observed in the offspring of women treated with DES. Such observations are verified by research based on the so-called relay methodology, based on the fact that methods of analysis for a residue and its metabolites fail to reflect practical realities, since they may lead to findings of residues without bioavailability or vice versa. Further, the individual biological evaluation of each metabolite fails to reflect the overall consumption of all of them. The relay, a farm animal, changes this situation.

In order to determine whether this was true of the first two types of anabolic agents, rats and mice of both sexes received balanced diets containing 20% of the flesh or 6% of the liver of male and female calves implanted with DES (24 mg in each of two implants), oestradiol and progesterone (20 + 200 mg in two doses) or oestradiol + testosterone (20 + 200 mg in two doses). Animals were implanted at the beginning of the growth period and 60 days later, 38 days prior to slaughter.

Chemical measurement of residues showed DES levels of 30 g/kg in the flesh, but none in the liver. Levels of natural steroid hormones did not exceed levels in untreated animals.

> *Feeding rats and mice on the flesh of DES-implanted calves led to growth in both sexes, sterilized the females and sharply atrophied the testicles, seminal ducts and penis of the males. Liver behaviour was less clear. Pokrovski et al. and Nesterin obtained similar results with flesh. It was also observed that DES persisted for nearly 40 days in the environment and was found in alfalfa.*

Animals fed on the flesh and liver of calves implanted with oestradiol + progesterone or oestradiol + testosterone behaved in the same manner as controls fed on the flesh and liver of control calves.

After 24 months, the percentage of tumors (primarily mammary adenomata, frequent in old Wistar rats) was the same in both groups.

> *Trenbolone acetate (TBA), a testosterone derivative, leaves little or no residue. Measured by radio-immunoassay, Gropp et al. also studied their biological effects by the relay toxicity method, on Sprague-Dawley rats fed on the flesh of calves implanted with a mixture of 20 mg oestradiol-17β + 140 mg TBA, as well as with doses 10 and 25 times higher. The results of this study, which continued for 110 weeks and covered two generations, showed the absence of any unfavourable influence on growth, organ weight, reproduction, teratogenicity and cancerogenicity. Only females fed on the flesh of control calves to which TBA had been added in doses 25 times normal, had significantly lower growth($P < 0.05$). The biochemical constants in the blood were the same in both groups.*

These experiments show that compounds of natural steroid male and female hormones, and their derivatives, appear to be without danger for the consumer. They are useful in animal production, obviating a number of often costly and polluting therapeutic operations. On the contrary, the use of DES-type non-steroid anabolic agents has serious disadvantages. It is difficult to detect their residues in meat. Relay toxicity results show that analyses and biological behaviour fail to agree.

The indiscriminate prohibition of all types of anabolic agents may be leading to the use of DES and similar substances. Analyses of urine in which they can be identified most easily, have revealed them in 20% of cases. In the United States, McClung estimates that 500 000 bovines are being treated illegally.

In Europe, illegal use appears to be fairly general, but is tending to decrease. The use of DES and similar substances should thus be prohibited, while that of the steroids and their derivatives should be authorized under veterinary control. This was the view of many writers as early as 1967, and their conclusions were confirmed by the FAO/WHO Joint Committee of 1975 and a recent international symposium (Warsaw, 27–30 April 1980). Any new anabolic agent should be severely tested before it is distributed widely.

There is a need to develop practical, economic and rapid — but very exact — methods of detecting residues, particularly those originating in products chemically different from the natural steroid hormones and their derivatives.

It should be clearly understood that this discussion concerns only measurement methods intended to detect residues remaining in flesh and viscera. These methods are, however, not very reliable, and other biological material must be examined to determine whether the animals have been treated. Specialists agree that there are still difficulties in testing for residues in the tissues of animals intended for use as food.

Many different methods have been proposed and discussed at the EEC level. While the trend is increasingly toward the use of radio-immunoassay, it is undeniable that the techniques are highly complex and still far from being perfected.

DES and similar substances should preferably be looked for in urine and faeces. Instructions to this effect were issued in the United States in 1979 and confirmed in 1980.

Account should also be taken of the fact that no difference may be found between animals properly implanted with anabolic agents and untreated animals.

This problem has received insufficient attention. The proceedings of the FAO/WHO Joint Meeting of March 1975 contain reports emphasizing this lack of difference and the relatively high, but physiological, levels of testosterone and progesterone in the flesh of bulls and gestating cows, respectively. This being the case, how can the natural presence of these substances be distinguished from their presence due to the use of natural products?

Further, it should not be forgotten that very many animal feeds contain phytohormones which can cause modifications in the morphology and histology of certain organs.

Two new methods may also be mentioned: high-pressure liquid chromatography which, according to Richou-Bac and Pantaleon does not appear to be sufficiently sensitive, and the so-called Elisa method, applied by Ruitenberg et al. in the Netherlands to detect pig infection by Trichinella spiralis.

This method, which uses an enzyme for labelling, followed by measurement with 450 nm spectrophotometry, needs further development, but it may be possible to automatize its use. Kroes, working at Bilthoven (Netherlands) is studying how it can be used to measure hormone levels.

In practice, the most urgent problem appears to be the detection of DES. Bories *et al.* and other authors have shown that it is eliminated in the urine and faeces. It has also been demonstrated that the saliva does not contain DES but that it does contain progesterone. DES cannot be detected in the flesh of animals treated three weeks before slaughter. It can be detected in the urine after one month by thin-layer chromatography and radio-immunoassay.

The bladder need only contain from 20 to 50 ml of urine for the thin-layer chromatographic method to be applied. The method, used in 1973 to detect oestradiol residues, was modified and improved by Verbeke in 1979. Vogt and Oehrle have employed it for detecting steroid oestrogen residues and Stilbene in calves' urine. Ryan and Hoffmann use it, together with radio-immunoassay, to study linked trenbolone residues.

For the latter compound, according to Richou-Bac and Pantaleon, Schuller and Stephany believe it possible to use thin-layer chromatography to attain a sensitivity level of 0.005 ppb, comparable to that for radioimmunoassay; this figure has not been included.

The hormone is rendered antigenic by localization on seric albumen. Labelling the hormone should in no case alter the molecule. The specificity of immunoassay depends on the intrinsic characteristics of the antibody and the efficiency of the hormone purification method. Many techniques exist for separating the free H*, but immunoprecipitation appears to be the method most commonly used. The immunological behaviour of the hormone present in its matrix should be identical with that of the reference standard.

It is not yet clear what techniques are the most exact and economical and are therefore practical for large-scale control.

The basic material which can be inspected also needs to be determined.

Flesh, fat and viscera cannot be used for DES, which can be identified only in the urine, with a sensitivity of about 6 ppb. Many analyses can be made by thin-layer chromatography and radio-immunoassay. The French control services use only the former, estimating the cost of one analysis at 25 FFr (= approx. US $5). The possibility of techniques applicable to flesh is under study, but in this regard the techniques speak of "chemical torture". They concede that routine control of meat for all hormones appears to be difficult.

The same services also use urine for measuring natural hormones, applying Verbeke's technique. No results are obtained from flesh, even for natural hormones, over 3 weeks following treatment, levels being too low. Thus, as compared with untreated calves, thin-layer chromatography yields no information. The various hormones, according to the solvent system used, as determined by Verbeke, and also the RF values and spot colours under UV fluorescence at 366 nm.

It can be concluded that routine control in flesh is very difficult in all cases, whatever the anabolic agent. Above all, the use of DES should be avoided, and for this purpose, using urine as the basic material, high performance bidimensional thin-layer chromatography is the method to be preferred.

Bibliography

Ahlgren, G.H.: *Forage Crops,* McGraw-Hill, New York, 2006.

Astley Maberly, C. T.: *Animals of East Africa,* Hodder & Stoughton, London, 2007.

Boehrer, Bruce: *A Cultural History of Animals in the Renaissance,* Oxford, UK: Berg Publishers, 2007.

Bogdan, A.V.: *Tropical Pasture and Fodder Legumes,* Longmans, London, 2003.

Bright, M.: *Animal Language,* BBC Publications, London, 2004.

Carl Cohen and Tom Regan: *The Animal Rights Debate,* Rowman & Littlefield, Lanham, MD, 2001.

Cheeke, P. R.: *Applied Animal Nutrition: Feeds and Feeding,* McGraw-Hill, New York, 2008

Church, D. C.: *Livestock Feeds and Feeding,* McGraw-Hill, New York, 2007

Cloudsley-Thompson, J. L.: *The Zoology of Tropical Africa,* W. W. Norton, New York, 2009.

Cullison, A. and R. S. Lowrey, *Feeds and Feeding,* McGraw-Hill, New York, 2006.

Curtis, H. and N. S. Barnes: *Biology,* Worth Publishers, Inc. New York, 2009.

Davidson, Peter Hobley. George Orwell: *A Literary Life,* St. Martin's Press, New York, 2006.

Dolins, Francine: *Attitudes to Animals: Views on Animal Welfare,* Cambridge University Press, Cambridge, 2009.

Ensminger, J. E. Oldfield, and W. W. Heinemann, *Feeds and Nutrition,* McGraw-Hill, New York, 2008

Ensminger, R. M.: *The Stockman's Handbook,* McGraw-Hill, New York, 2012.

Estes, Richard Despard: *The Behaviour Guide to African Animals,* University of California Press, Berkeley, CA, 2011.

Fairey, D.T., & Hampton, J.G.: *Forage Seed Production of Temperate Species,* CAB, Farnham Royal, 2007.

Fogle, Bruce: *Pets and Their People,* The Viking Press, New York, 1983.

Fox, Michael Allen: *The Case for Animal Experimentation,* University of California Press, Berkeley, CA, 2006.

Frame, J., Charlton, J.F.L., & Laidlaw, A.S.: *Temperate Forage Legumes,* CAB Interational, Wallingford, 2008.

Frame, J.: *Improved Grassland Management,* Farming Press, Ipswich, 2012.

Frederick, Zeuner E.: *A History of Domesticated Animals,* Hutchinson, London, 2003.

Friend, Tim: *Animal Talk: Breaking the Codes of Animal Language,* Free Press, New York, 2004.

Fudge, Erica: *Brutal Reasoning: Animals, Rationality and Humanity in Early Modern England,* Cornell University Press, Ithaca, 2006.

Gallistel, C.R.: *Animal Cognition,* MIT Press, Cambridge, 2012.

Gates, P.: *Animal Communication,* Cambridge University Press, Cambridge, 2007.

Giorgio, Agamben: *The Open: Man and Animal,* Stanford University Press, UK, 2004.

Hacker, J.B.: *Nutritional Limits to Animal Production from Pasture,* CAB, Farnham Royal, 2011.

Hanson, A.A., Barnes, D.K., & Hill, R.R.: *Alfalfa and Alfalfa Management,* Wisconsin, Madison, 2008.

Hanson, C.H. : *Alfalfa Science and Technology,* Madison, Wisconsin, 1998.

Harris, Marvin: *The Sacred Cow and the Abominable Pig: Riddles of Food and Culture,* Touchstone Books, New York, 2007.

Hearne, Vicki: *Animal Happiness,* HarperCollins, New York, 1994.

Hitchcock, A.S., & Chase, A.: *Manual of Grasses of the United States,* USDA, 2011.

Holmes, W.: *Grass, its Production and Utilisation,* Blackwell, Oxford & London, 2009.

Iwago, Mitsuaki: *Serengeti: Natural Order on the African Plain,* Chronicle Books,San Francisco, CA, 2009.

Jones, M.B. & Lazenby, A.: *The Grass Crop,* Chapman & Hall, London, 2008.

Julian Baldick: *Animals and Shaman: Ancient Religions of Central Asia,* New York University Press, New York, 2000.

Juliet: *Domesticated Animals from Early Times,* University of Texas Press, Austin, TX, 2011.

Karen, Allen Miller: *The Human-Animal Bond: An Annotated Bibliography,* Scarecrow Press, Metuchen, NJ, 2005.

Kevin Dolan: *Ethics, Animals, and Science,* Blackwell Science, Malden, MA, 2009.

Kiss, Agnes: *Living with Wildlife: Wildlife Resource Management with Local Participation in Africa,* World Bank, Washington, D.C., c1990.

Langley , Gill: *Animal Experimentation: The Consensus Changes*, Chapman and Hall, New York, 2009.

Lawick, Hugo van: *Among Predators and Prey*, Sierra Club Books, San Francisco, 2006.

Lu, F.C. & Rendel, J.: *Anabolic Agents in Animal Production,* FAO/WHO Symposium, Rome, 2005.

Mason, Jim and Peter Singer: *Animal Factories*, Crown Publishers, New York, 2010.

Masson, Jeffrey Moussaieff: *Dogs Never Lie About Love: Reflections on the Emotional World of Dogs*, Crown Publishers, New York, 2007.

McDonald, P., Edwards, R.A., Grenhalgh, J.F.D., & Morgan, C.A.: *Animal Nutrition,* Longmans, London, 2005.

Morton, Eugene S.: *Animal Talk: Science and the Voices of Nature,* Random House, New York, 2012.

Natz, D.: *Feed Additive Compendium*, McGraw-Hill, New York, 2007.

O'Neill, Terry: *Readings on Animal Farm*, Greenhaven Press, San Diego, 2009..

Orwell, George: *Animal Farm: A Fairy Story*, Signet Classics, Orlando, 2009.

Pietro Croce: *Vivisection or Science?: An Investigation into Testing Drugs and Safeguarding Health*, Zed Books, New York, 1999.

Pond, W. G.: *Basic Animal Nutrition and Feeding*, McGraw-Hill, New York,, 2005.

Randall, D., W. Burggren, and K. French: *Eckert Animal Physiology*, W.H. Freeman and Company, New York, 2007.

Raymond, F., Rēdman, P., & Waltham, R.: *Forage conservation and Feeding,* Farming Press, Ipswich, 2006.

Roger, J.: *Buffon: A Life in Natural History,* Cornell University Press, Ithaca, NY, 1997.

Roy, J.H.B.: *Studies in the Agricultural and Food Sciences: the Calf*, Butterworths, London, 2010.

Ruth Ellen Bulger: *The Ethical Dimensions of the Biological and Health Sciences*, Cambridge University Press, New York, 2002.

Service, Robert: *A History of Modern Russia*, Harvard University Press, Cambridge, Mass, 2005.

Sharpe, Robert: *Science on Trial: The Human Cost of Animal Experiments*, Awareness Books, Sheffield, UK, 2004.

Skerman, P.J., & Riveros, F.: *Tropical Grasses,* FAO, Rome, 2010.

Skerman, P.J., Cameron, D.G., & Riveros, F.: *Tropical Forage Legumes,* FAO, Rome, 2011.

Snaydon, R.W.: *Managed Grasslands,* Elsevier, Amsterdam & London, 2013.

Steve Baker: *The Postmodern Animal,* Reaktion Books, London, 2000.

Stuart, Chris and Tilde Stuart: *Africa's Vanishing Wildlife,* Smithsonian Institution Press, Washington, D.C., 2006.

Taylor, R.E. and R. Bogart: *Scientific Farm Animal Production: An Introduction to Animal Science,* MacMillan, New York, 2009.

Ted Benton: *Natural Relations: Ecology, Animal Rights and Social Justice,* Verso, London, 2013.

Umali, D. & Schwartz, L.: *Public and Private Agricultural Extension: Beyond Traditional Frontiers,* Washington, DC, World Bank, 2004.

VanderWal, P.: *Anabolic Agents in Animal Production,* Environmental Quality and Safety, Suppl. 2006.

Walton, P.D.: *Production and Management of Cultivated Fodders,* Reston Publishing, Reston, VA, 2012.

Whyte, R.O, Moir, T.R.G., & Cooper, J.P.: *Grasses in Agriculture,* FAO, Rome, 2009.

Whyte, R.O., Nillson-Leissner, G., & Trumble, H.C.: *Legumes in Agriculture,* FAO, Rome, 2003.

Wieczynski, Joseph L.: *The Modern Encyclopedia of Russian and Soviet History,* Academic International Press, Gulf Breeze, Fla, 2006.

Willis, R.G.: *Signifying Animals: Human Meaning in the Natural World,* Unwin Hyman, London, 2010.

Yarri, Donna: *The Ethics of Animal Experimentation: A Critical Analysis And Constructive Christian Proposal,* Oxford University Press, Oxford, 2005.

Index

□□□